T0137265

Spatially Explicit Hyperparameter Optimization for Neural Networks

Minrui Zheng

Spatially Explicit Hyperparameter Optimization for Neural Networks

Springer

Minrui Zheng
School of Public Administration and Policy
Renmin University of China
Beijing, China

ISBN 978-981-16-5401-5 ISBN 978-981-16-5399-5 (eBook)
https://doi.org/10.1007/978-981-16-5399-5

This Springer imprint is published by the registered company Springer Nature Singapore Pte Ltd.
The registered company address is: 152 Beach Road, #21-01/04 Gateway East, Singapore 189721, Singapore

To my parents and my advisor (Dr. Wenwu Tang). Their endless support has been encouraging to continue my research.

Preface

Neural networks as a commonly used machine learning algorithms (e.g., artificial neural networks and convolutional neural networks) have been extensively used in GIScience domain to explore the nonlinear or/and complex geographic phenomena. However, how to automatically adjust the parameters of neural networks is still an open question in GIScience. Moreover, the model performance of neural networks often depends on the parameter setting for a given dataset. Meanwhile, adjusting the parameter configuration of neural networks will increase the overall running time. In this book, the author proposed an automated spatially explicit hyperparameter optimization approach to identify optimal or near-optimal parameter settings for neural networks and accelerate the search process through both model and computing levels. The author used two spatial prediction models in this book to examine the utilities of spatially explicit hyperparameter optimization. The results demonstrate that the approach proposed in this book improves the computing performance at model and computing levels and addresses the challenge of finding optimal parameter settings for neural networks in the GIScience field.

In the remainder of this book, Chap. 2 focuses on a literature review of artificial neural networks, hyperparameter optimization, cyberinfrastructure and high-performance and parallel computing, and evolutionary algorithms. Chapter 3 describes the framework of spatially explicit hyperparameter optimization. Chapter 4, which connects to objective 1, focuses on introducing the basic framework of spatially explicit hyperparameter optimization that incorporates spatial statistics and high-performance and parallel computing. Chapter 5 demonstrates the utilities of the automated spatially explicit hyperparameter optimization (objective 2). Chapter 6 examines the practicability of spatially explicit hyperparameter optimization, which links to objective 3. Chapter 7 concludes this book.

Beijing, China — Minrui Zheng

Acknowledgements

My deepest gratitude goes to my committee members, Drs. Wenwu Tang, Elizebeth Delmelle, Minwoo Lee, and Akin Ogundiran for their support and guidance on this work.

I owe many thanks to former and current members of Center for Applied GIScience (Dr. Michael Desjardins, Dr. Alexander Hohl, Yu Lan, Dr. Jianxin Yang, Tianyang Chen, Zachery Slocum) and faculty (Drs. Heather Smith, Eric Delmelle, Craig Allan, Yu Wang, and Lisa Russell-Pinson) at the University of North Carolina at Charlotte who have helped and encouraged me. Also, I would like to express my appreciation to my friends (Greg Verret, Amanda Verret, Mark Verret, Nathan Verret, Yi Zhang, Li Liu, Qiang Li, and Jiayang Li).

Contents

List of Figures

List of Tables

Chapter 1
Introduction

1.1 Background

In the past decades, with the increasing volume of spatial data and development of cutting-edge techniques, several spatial models have been created to investigate complex spatial phenoména and explore spatial process (Goodchild 1992; Longley and Batty 1996; Graham 1997; Fotheringham et al. 2000; Miller and Goodchild 2015).

Spatial modeling embraces a series of models and techniques that explore relationships, patterns, and phenomena across space and time. The steps of spatial modeling often proceed in a sequence from problem specification, model theory, data preparation, model verification, calibration, and evaluation to prediction (see Fig. 1.1). Although spatial modeling traditionally belongs to the domain of geography, it can be applied to a variety of geography-related domains, e.g., ecology, urban studies, transportation, and social science (Longley and Batty 1996; Miller 1999; Krewski et al. 2009; Borcard et al. 2011; Logan 2012), as well as other disciplinary domains, including computer science and mathematics (Gelfand et al. 2005; Andrews et al. 2010).

Although spatial modeling exists in a number of research areas, one of the vital parts of spatial modeling is algorithms. Figure 1.2 illustrates the process of algorithms from input X to output Y. Algorithms are a sequence of computational steps that transform the input into the output. In each spatial modeling exercise, it includes one or multiple model units which have a single algorithm and related parameters. There are two types of parameters based on their contributions to spatial modeling: standard parameters and hyperparameters. Standard parameters are an "internal" component of spatial modeling, and their values usually are derived from models, such as coefficients of regression models and coefficients of objective functions of optimization models.

Hyperparameters are an "external" component of spatial modeling, and the values are user-defined or pre-defined by other algorithms. Hyperparameters usually influence the algorithms themselves and the derivation of standard parameters. Some

M. Zheng, *Spatially Explicit Hyperparameter Optimization for Neural Networks*,
https://doi.org/10.1007/978-981-16-5399-5_1

Fig. 1.1 Spatial modeling process (Adapted from Shannon 1975 and Batty 1976)

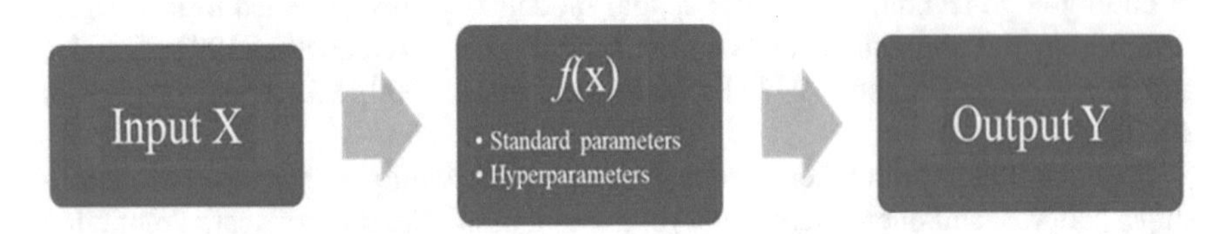

Fig. 1.2 Illustration of algorithms (Y = f(X); Y = [y_1, y_2,..., y_n]; X = [x_1, x_2,..., x_m]; f is the algorithm between X and Y), revised from Gahegan (2003)

examples of hyperparameters are the learning rate of artificial neural networks (ANNs), the initial number of clusters of k-means clustering, and the number of trees in random forest algorithm. However, hyperparameters also exist in rule-based spatial modeling, such as the cellular automata (CA) model. Stochastic disturbance term from transition rules is a hyperparameter for CA, which allows the CA model generated patterns to be closed to reality (Yeh and Li 2001).

With consideration of big spatial data and complex spatial phenomena (Couclelis 1998; Longley et al. 1998; Openshaw and Abrahart 2000), the emergence of GeoComputation opened a new computational world to geographers because it provides more computational methods and techniques that are applicable to geographical problems (Longley et al. 1998; Couclelis 1998; Openshaw and Abrahart 2014; Brunsdon and Singleton 2015). GeoComputation has four core aspects (Gahegan 1999): computer architecture and design (e.g., parallel computing); search, classification, prediction and modeling (e.g., artificial neural network); knowledge discovery (e.g., data mining); and visualization (e.g., statistical results using graphics).

Machine learning belongs to the second aspect of GeoComputation (search, classification, prediction, and modeling). A number of geographers are placing emphasis on the use of machine learning algorithms for spatial modeling (Batty et al. 1999; Sui and Maggio 1999; Pijanowski et al. 2002, 2014; Li and Yeh 2002). But, several studies discussed hyperparameter settings for machine learning algorithms. Finding optimal hyperparameters thus can often make the difference between average results and state-of-the-art performance. Most of hyperparameter optimization-related articles were published in computer science and engineering (Chapelle et al. 2002; Bergstra and Bengio 2012; Bergstra et al. 2013a; Thornton et al. 2013). In geography field, the most commonly used hyperparameter optimization approach is manual selection (Li and Yeh 2002; Pijanowski et al. 2005, 2014; Zheng et al. 2019), although a number of automated hyperparameter optimization approaches and applications were proposed by researchers from computer science (Bergstra et al. 2013b; Thornton et al. 2013; Li et al. 2017).

Hyperparameter optimization has data- and computational-intensity issues, which have been identified by current hyperparameter optimization studies (Bergstra et al. 2013a; Bergstra et al. 2015; Lorenzo et al. 2017; Falkner et al. 2018). In addition, as the volume of spatial data increases, the process of finding optimal or near-optimal hyperparameters is becoming a computational-intensive problem. In order to address these limitations, cyberinfrastructure (CI) and high-performance and parallel computing (HPC) provide a solution. CI is "a combination of data resources, network protocols, computing platforms, and computational services that bring people, information, and computational tools together to perform science or other data-rich applications in this information-driven world" (Page 1) (Yang et al. 2010). HPC includes grid computing, cluster computing, and ubiquitous computing, which provides super computing power for CI's applications (Yang et al. 2010; Tang et al. 2018). Existing hyperparameter optimization approaches or applications typically adopted HPC to address the computational bottleneck (Bergstra et al. 2015; Falkner et al. 2018). Meanwhile, the capabilities and importance of CI and HPC in spatial modeling have been discussed in a series of studies (Armstrong 2000; Wang and Liu 2009; Yang et al. 2010; Wang 2010; Tang et al. 2018).

However, the discussions of hyperparameter optimization-related studies are insufficient. Also, current hyperparameter optimization approaches assumed hyperparameters are independent. According to the First Law of Geography—"everything is related to everything else, but near things are more related than distant things" (Tobler 1970), similar things are placed closer than dissimilar things. The "distance" between two things is not only about the distance in real-world (e.g., spatial units (meters or miles) or traveling time), it also can be measured as similarity in a virtual space (Waters 2017). Thus, the assumption of current hyperparameter optimization is inappropriate. Since each combination of hyperparameters may contain information relative to others, the methods from geography may help to identify the similarity (spatial dependence) and explore the landscape of the hyperparameter space. Hence, a hyperparameter approach that considers the similarity and landscape of hyperparameter space is necessary, particularly within the context of spatial modeling.

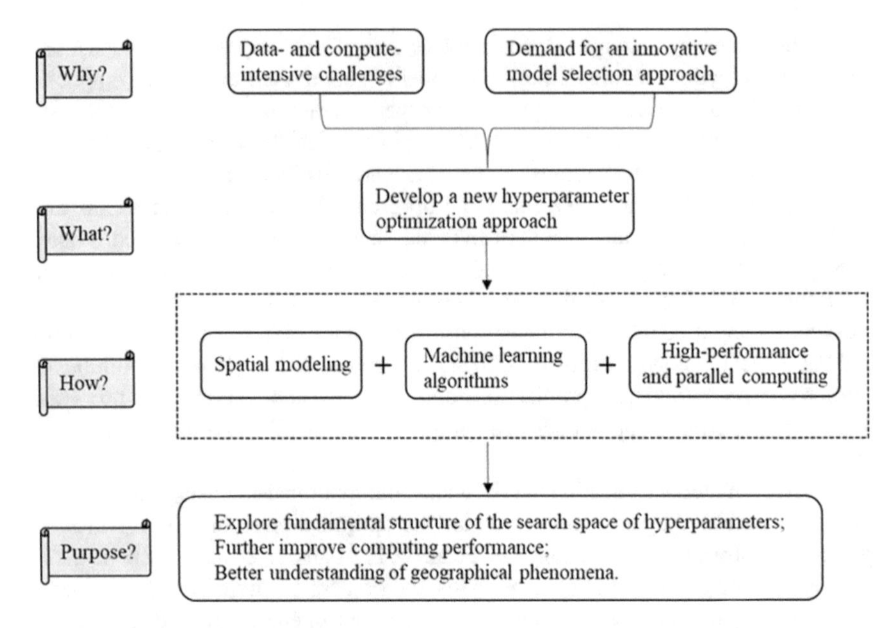

Fig. 1.3 Roadmap of this book

Figure 1.3 is the roadmap of this book showing the motivations, the methods, and the contributions. In this book, I propose a new hyperparameter optimization approach that is closely related to spatial modeling, machine learning algorithms, and HPC. The specific research objectives will be discussed in Sect. 1.2.

1.2 Research Objectives

There are a number of machine learning algorithms that are applied in spatial modeling, such as artificial neural networks (ANNs), support vector machines (SVMs), genetic algorithms (GAs), and random forests (RFs). As one of the popular machine learning algorithms, ANNs are a practical and widely applicable algorithm in current geography-related research (Zheng et al. 2019). A series of neural network-related studies have been carried out on various aspects of geography-related fields, such as land change models, remote sensing, and urban planning (Li and Yeh 2002; Pijanowski et al. 2002; Nevtipilova et al. 2014). Hence, I examine the utility of spatially explicit hyperparameter optimization approach in neural network-based spatial models in this book.

While ANNs have been the subject of many classic studies in GIScience[1] field, the search process of hyperparameter optimization for ANN-based spatial models is still a "black-box" or "gray-box" problem. The performance of ANNs depends crucially on the specific hyperparameters, such as the learning rate. The quality and features of data may influence the hyperparameters as well. Most previous studies used trial-and-error approach or expert opinions (Li and Yeh 2002; Pijanowski et al. 2005, 2014). These methods are easy to use, but, they are subjective. Moreover, ANNs are a data-driven method, different datasets may have different hyperparameter settings. Thus, finding optimal or near-optimal hyperparameters of ANNs can often make the difference between average results and state-of-the-art performance.

In this book, I present a spatially explicit hyperparameter optimization approach, which is an improvement for existing approaches by taking into account spatial characteristics (i.e., spatial dependence). In addition, I develop an automated framework of spatially explicit hyperparameter optimization for ANN-based spatial models, which allows for (1) handling the data- and computational-intensity challenges of hyperparameter optimization, using cyber-enabled HPC, (2) handling multiple hyperparameters simultaneously, and (3) explore the local variation structure of hyperparameters. Lastly, I apply the proposed spatially explicit hyperparameter optimization in different case studies. The results of case studies might guide practitioners on improving their model performance. Therefore, I use three research objectives for advancing the body of knowledge:

1.2.1 Objective 1

- Examine the feasibility and necessity of hyperparameter optimization in machine learning algorithm-based spatial models;
- Evaluate how the methods from geography may help to identify the similarity and explore the landscape of the hyperparameter space;
- Explore the potential feasibility of accelerating hyperparameter optimization from the model-level.

1.2.2 Objective 2

- Automate the spatially explicit hyperparameter optimization approach that considers spatial dependence in the optimization process;
- Examine the utility of the automated spatially explicit hyperparameter optimization;

[1] "Geographic information science (GIScience), which is the research field that studies the general principles underlying the acquisition, management, processing, analysis, visualization, and storage of geographic data" (page 494) (Goodchild 2003).

- Explore the local variation structure of the search space of hyperparameters and adjust the local variation structure based on spatial dependence.

1.2.3 Objective 3

- Investigate the universality of spatially explicit hyperparameter optimization in different neural network-based spatial models;
- Evaluate the model performance and efficiency of spatially explicit hyperparameter optimization;
- Identify future directions of spatially explicit hyperparameter optimization.

References

Andrews, Jeffrey G, Radha Krishna Ganti, Martin Haenggi, Nihar Jindal, and Steven Weber. 2010. A primer on spatial modeling and analysis in wireless networks. *IEEE Communications Magazine* 48 (11).

Armstrong, Marc P. 2000. Geography and computational science.

Batty, M. 1976. Urban modelling; algorithms, calibrations, predictions.

Batty, Michael, Yichun Xie, and Zhanli Sun. 1999. Modeling urban dynamics through GIS-based cellular automata. *Computers, Environment and Urban Systems* 23 (3): 205–233.

Bergstra, James, and Yoshua Bengio. 2012. Random search for hyper-parameter optimization. *Journal of Machine Learning Research* 13: 281–305.

Bergstra, James, Brent Komer, Chris Eliasmith, Dan Yamins, and David D Cox. 2015. Hyperopt: A python library for model selection and hyperparameter optimization. *Computational Science & Discovery* 8 (1):014008.

Bergstra, James, Dan Yamins, and David D Cox. 2013a. Hyperopt: A python library for optimizing the hyperparameters of machine learning algorithms. In *Proceedings of the 12th python in science conference.*

Bergstra, James, Daniel Yamins, and David D. Cox. 2013b. Making a science of model search: Hyperparameter optimization in hundreds of dimensions for vision architectures. *ICML* 1 (28): 115–123.

Borcard, Daniel, François Gillet, and Pierre Legendre. 2011. Spatial analysis of ecological data. In *Numerical ecology with R*, 227–292. Springer.

Brunsdon, Chris, and Alex Singleton. 2015. *Geocomputation: A practical primer*. Sage.

Chapelle, Olivier, Vladimir Vapnik, Olivier Bousquet, and Sayan Mukherjee. 2002. Choosing multiple parameters for support vector machines. *Machine Learning* 46 (1–3): 131–159.

Couclelis, Helen. 1998. Geocomputation in context. Geocomputation: A primer.

Falkner, Stefan, Aaron Klein, and Frank Hutter. 2018. BOHB: Robust and efficient hyperparameter optimization at scale.

Fotheringham, A Stewart, Chris Brunsdon, and Martin Charlton. 2000. *Quantitative geography: Perspectives on spatial data analysis*. Sage.

Gahegan, Mark. 1999. Four barriers to the development of effective exploratory visualisation tools for the geosciences. *International Journal of Geographical Information Science* 13 (4): 289–309.

Gahegan, Mark. 2003. Is inductive machine learning just another wild goose (or might it lay the golden egg)? *International Journal of Geographical Information Science* 17 (1): 69–92.

Gelfand, Alan E., Athanasios Kottas, and Steven N. MacEachern. 2005. Bayesian nonparametric spatial modeling with Dirichlet process mixing. *Journal of the American Statistical Association* 100 (471): 1021–1035.

Goodchild, Michael F. 1992. Computers, and Geosciences. *Geographical Data Modeling* 18 (4): 401–408.

Goodchild, Michael F. 2003. Geographic information science and systems for environmental management. *Annual Review of Environment and Resources* 28.

Graham, E. 1997. Philosophies underlying human geography research. In *Methods in human geography*, eds. R. Flowerdew and D. Martin D. Harlow: Longman.

Krewski, Daniel, Michael Jerrett, Richard T Burnett, Renjun Ma, Edward Hughes, Yuanli Shi, Michelle C Turner, C Arden Pope III, George Thurston, and Eugenia E Calle. 2009. *Extended follow-up and spatial analysis of the American Cancer Society study linking particulate air pollution and mortality*. Boston, MA: Health Effects Institute.

Li, Lisha, Kevin Jamieson, Giulia DeSalvo, Afshin Rostamizadeh, and Ameet Talwalkar. 2017. Hyperband: A novel bandit-based approach to hyperparameter optimization. *The Journal of Machine Learning Research* 18 (1): 6765–6816.

Li, Xia, and Anthony Gar-On Yeh. 2002. Neural-network-based cellular automata for simulating multiple land use changes using GIS. *International Journal of Geographical Information Science* 16 (4): 323–343.

Logan, John R. 2012. Making a place for space: Spatial thinking in social science. *Annual Review of Sociology* 38: 507–524.

Longley, Paul A, and Michael Batty. 1996. *Spatial analysis: Modelling in a GIS environment*. Wiley.

Longley, Paul A, Susan Brooks, W Macmillan, and RA McDonnell. 1998. *Geocomputation: A primer*. Wiley.

Lorenzo, Pablo Ribalta, Jakub Nalepa, Luciano Sanchez Ramos, and José Ranilla Pastor. 2017. Hyper-parameter selection in deep neural networks using parallel particle swarm optimization. In *Proceedings of the genetic and evolutionary computation conference companion*.

Miller, Harvey J. 1999. Potential contributions of spatial analysis to geographic information systems for transportation (GIS-T). *Geographical Analysis* 31 (4): 373–399.

Miller, Harvey J, and Michael F Goodchild. 2015. Data-driven geography. *GeoJournal* 80(4): 449–461.

Nevtipilova, Veronika, Justyna Pastwa, Mukesh Singh Boori, and Vit Vozenilek. 2014. Testing artificial neural network (ANN) for spatial interpolation. *International Journal of Geology and Geosciences (JGG)* 01–09. ISSN 2329 6755.

Openshaw, Stan, and RJ Abrahart. 2014. *GeoComputation*, 2nd edn, 1–21. Boca Raton: CRC Press.

Openshaw, Stan, and Robert J Abrahart. 2000. *GeoComputation*, vol. 24. London: Taylor & Francis.

Pijanowski, Bryan C., Daniel G. Brown, Bradley A. Shellito, and Gaurav A. Manik. 2002. Using neural networks and GIS to forecast land use changes: A land transformation model. *Computers, Environment and Urban Systems* 26 (6): 553–575.

Pijanowski, Bryan C., Snehal Pithadia, Bradley A. Shellito, and Konstantinos Alexandridis. 2005. Calibrating a neural network-based urban change model for two metropolitan areas of the Upper Midwest of the United States. *International Journal of Geographical Information Science* 19 (2): 197–215.

Pijanowski, Bryan C., Amin Tayyebi, Jarrod Doucette, Burak K. Pekin, David Braun, and James Plourde. 2014. A big data urban growth simulation at a national scale: Configuring the GIS and neural network based land transformation model to run in a High Performance Computing (HPC) environment. *Environmental Modelling & Software* 51: 250–268.

Shannon, Robert E. 1975. Systems simulation; the art and science.

Sui, D.Z., and R.C. Maggio. 1999. Integrating GIS with hydrological modeling: Practices, problems, and prospects. *Computers, Environment and Urban Systems* 23 (1): 33–51.

Tang, Wenwu, Wenpeng Feng, Jing Deng, Meijuan Jia, and Huifang Zuo. 2018. Parallel Computing for Geocomputational Modeling. In *GeoComputational analysis and modeling of regional systems*, 37–54. Springer.

Thornton, Chris, Frank Hutter, Holger H Hoos, and Kevin Leyton-Brown. 2013. Auto-WEKA: Combined selection and hyperparameter optimization of classification algorithms. In *Proceedings of the 19th ACM SIGKDD international conference on knowledge discovery and data mining.*

Tobler, Waldo R. 1970. A computer movie simulating urban growth in the Detroit region. *Economic Geography* 46 (sup1): 234–240.

Wang, Shaowen. 2010. A CyberGIS framework for the synthesis of cyberinfrastructure, GIS, and spatial analysis. *Annals of the Association of American Geographers* 100 (3): 535–557.

Wang, Shaowen, and Yan Liu. 2009. TeraGrid GIScience gateway: Bridging cyberinfrastructure and GIScience. *International Journal of Geographical Information Science* 23 (5): 631–656.

Waters, Nigel. 2017. Tobler's first law of geography. https://doi.org/10.1002/9781118786352.wbieg1011.

Yang, Chaowei, Robert Raskin, Michael Goodchild, and Mark Gahegan. 2010. Geospatial cyberinfrastructure: Past, present and future. *Computers, Environment and Urban Systems* 34 (4): 264–277.

Yeh, Anthony Gar-On, and Xia Li. 2001. A constrained CA model for the simulation and planning of sustainable urban forms by using GIS. *Environment Planning B: Planning Design* 28(5): 733–753.

Zheng, Minrui, Wenwu Tang, and Xiang Zhao. 2019. Hyperparameter optimization of neural network-driven spatial models accelerated using cyber-enabled high-performance computing. *International Journal of Geographical Information Science.* https://doi.org/10.1080/13658816.2018.1530355.

Chapter 2
Literature Review

This literature review covers four topics, including artificial neural networks, hyperparameter optimization, cyberinfrastructure and high-performance and parallel computing, and evolutionary algorithms.

2.1 Artificial Neural Network

The first artificial neural network (i.e., perceptron) was designed in 1958 (Rosenblatt 1958). Since then, more and more studies adopt ANNs in their studies to capture nonlinear relationships between input and output, and solve complex real-world problems (Roberts and Attoh-Okine 1998; Li and Yeh 2002; Gulliford et al. 2004; Neaupane and Adhikari 2006). As identified by some pioneers from geography, ANNs and other machine learning algorithms can explore patterns and investigate complex relationships from spatial data with multiple scales and different spatial resolutions (Openshaw and Openshaw 1997; Gopal 2017). Further, ANNs and other machine learning algorithms are well suited to capturing nonlinearity from spatial data. Figure 2.1 shows the growth trend of the number of publications of ANN-based spatial modeling from 1990–2019. From this figure, we can see that the number of publications using ANNs in spatial modeling is not constantly growing since it was developed. The number of new publications kept a low or negative growth rate in the middle of the 1990s and early 21st century. Since 2014, the growth trend of research outputs (i.e., publications) has entered a stage of rapid growth. It is directly associated with the faster computing processors and new extensions of neural networks (e.g., convolutional neural networks) (Mellit and Kalogirou 2008; Niu et al. 2016).

The ability of ANNs in examining nonlinear relationships between input and output has been discussed in the past years (Hornik et al. 1989; Chen et al. 1990;

M. Zheng, *Spatially Explicit Hyperparameter Optimization for Neural Networks*,
https://doi.org/10.1007/978-981-16-5399-5_2

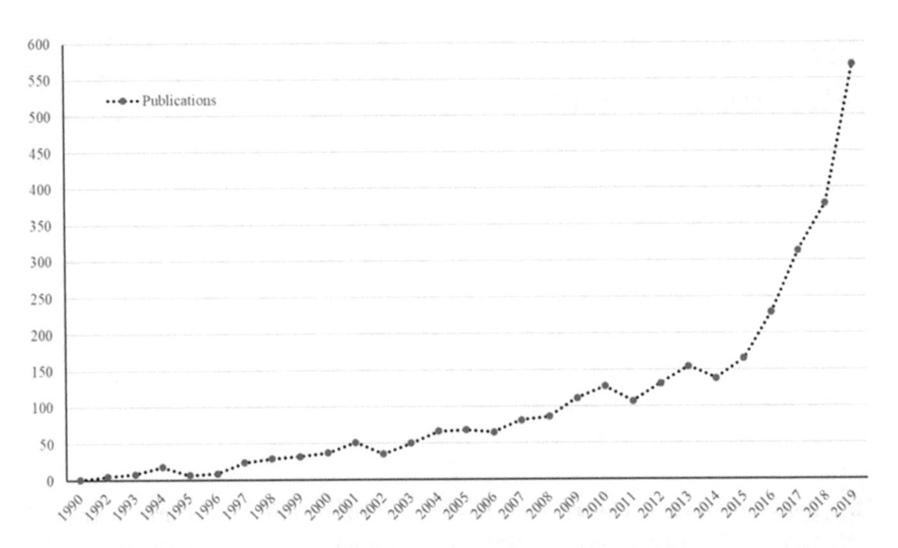

Fig. 2.1 Growth trend of spatial modeling and neural network-related publications from 1990–2019 based on Web of Science database (keywords: neural network and spatial modeling (or spatial analysis, or spatial analysis and modeling, or spatially explicit model, or geospatial analysis, or geospatial modeling); the total number of publication is 3,095)

Brondino and Silva 1999; Limsombunchai 2004; Mas and Flores 2008; Pijanowski et al. 2014). ANNs cover different types of neural networks, such as feedforward neural networks, recurrent neural networks, and convolutional neural networks. The latter is also an application of deep learning techniques. Input layer, hidden layer, and output layer are the basic types of layers used in traditional neural networks (Fig. 2.2).

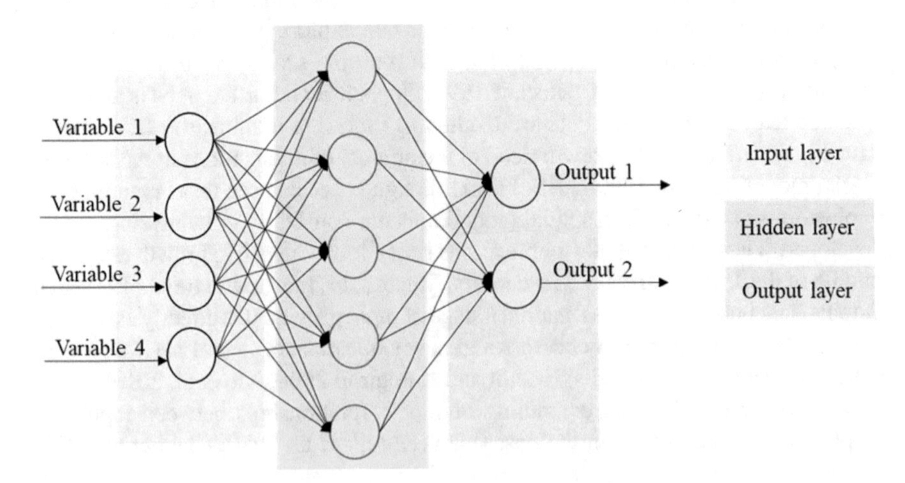

Fig. 2.2 Basic structure of a neural network with 4 input variables and 2 outcomes (network is fully connected)

Each input variable is linked to a node in the input layer, and each outcome is linked to a node in the output layer. Each connection (e.g., from input layer to hidden layer, from hidden layer to output layer) has a weight value associated with it. Learning algorithms are used to adjust the weights in order to minimize the error in the output. There are two types of learning algorithms, supervised learning algorithms (e.g., backpropagation) and unsupervised learning algorithms (e.g., self-organizing map). The goal of supervised learning is to map a function from input to desired output (Nasrabadi 2007). The goal of unsupervised learning is to explore the potential patterns and structures based on a series of data without pre-defined features or pre-defined objectives (Sanger 1989). Further, activation functions are used to determine the output based on a set of inputs. The major contribution of activation functions is to involve nonlinearity into neural networks. Other examples of activation functions include sigmoid, hyperbolic tangent, and rectified linear unit (ReLU) (Specht 1990; Hahnloser et al. 2000; Bishop 2006).

A number of neural network-based spatial models have been constructed in GIScience. For example, Atkinson and Tatnall used feedforward backpropagation multi-layer neural network to classify land cover types from Landsat Thematic Mapper imagery, and to identify and classify different categories of clouds (e.g., cirrus and stratocumulus) from Landsat Multispectral Scanner data (Atkinson and Tatnall 1997). They mentioned that neural networks provide more accurate results than traditional statistical approaches (e.g., maximum likelihood) and less computing time, and allow to classify images from multi-resource spatial data. Similarly, Xu et al. (2017) demonstrated the ability of convolutional neural networks to classify multi-resource remote sensing data. Li and Yeh (2002) simulated land use land cover changes using an integrated method (neural networks and cellular automata). The results showed that neural network-based cellular automata (CA) model can handle the multiple land use changes simulation, which is difficult for standard CA model. Based on Li and Yeh's work, a series of studies adopted neural network-based CA model in land change models or further refined this network-based CA model, such as Yeh and Li (2003), Pijanowski et al. (2005), Guan et al. (2005), Almeida et al. (2008), and Omrani et al. (2017).

Although there are many applications of neural network-based spatial models (Fig. 2.1), a few studies discussed the hyperparameters of neural networks (Wanas et al. 1998; Stathakis 2009; Karsoliya 2012). Besides the quality of datasets, the hyperparameters of neural networks have a significant influence on model performance. The investigations on how to find appropriate hyperparameters have been conducted in many years and the most remarkable contributions of existing hyperparameter optimization approaches came from computer science. Trial-and-error and recommendations from previous studies are the most popular method used in neural network-based spatial models. But, they have a common weakness–all of them are subjective (Zheng et al. 2019). Therefore, it is necessary to encourage spatial modelers to adopt hyperparameter optimization.

2.2 Hyperparameter Optimization

The importance of hyperparameters in neural networks has been discussed in the previous section, but how to effectively find appropriate hyperparameters is still a challenge. The process of determining appropriate hyperparameters is called hyperparameter optimization. Sampling can provide supports for accelerating the process of hyperparameter optimization. Grid search and random search are two widely used sampling methods in hyperparameter optimization. For grid search (Fig. 2.3A), we first need to define a grid size of the search space and then divide the entire search space into a number of grids based on the grid size. The hyperparameter at each grid point is a sampled hyperparameter. Random search is to randomly select sampled hyperparameters, and each combination of hyperparameters has an equal chance of being sampled (Fig. 2.3B). However, a random search may result in uneven distribution of samples. As the example in Fig. 2.3B shows, the distribution of samples is uneven; some areas have more samples than other areas. Latin hypercube sampling can resolve this limitation of random search. Latin hypercube sampling is a combination of grid search and random search. That is, we also need to separate the search space into a number of grids (Latin squares here), and sampled hyperparameters come from those Latin squares, and the combination of hyperparameters has an equal chance of being selected (Fig. 2.3C).

Bayesian optimization is another way to find optimal hyperparameters, which can find appropriate hyperparameters with less computing time (Snoek et al. 2012; Shahriari et al. 2015). Bayesian optimization (Močkus 1975) is a global optimization approach, which creates a model based on known sampled hyperparameters. Sequential model-based optimization (SMBO) also refers to Bayesian optimization. Typically, these known sampled hyperparameters are drawn from a Gaussian distribution. This step is called Gaussian process (GP), a stochastic method that is an extension of Gaussian probabilistic distribution (Williams and Rasmussen 2006). Acquisition function is used to create a utility function from these known sampled hyperparameters, which guides next search of hyperparameter optimization. Acquisition function is the key component of Bayesian optimization because the goal of

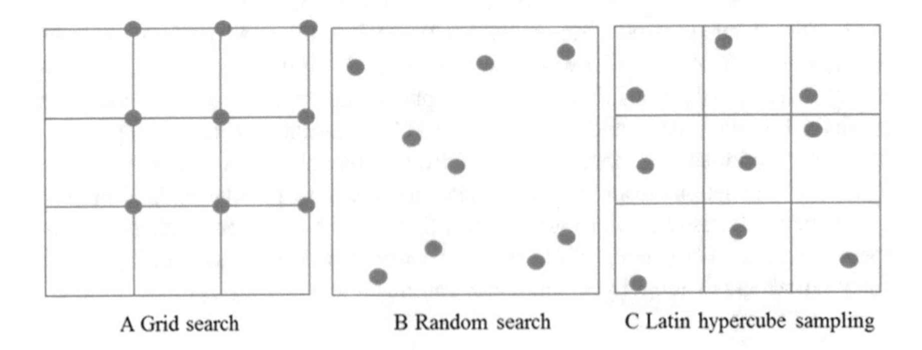

Fig. 2.3 Sampling methods with 9 sample points

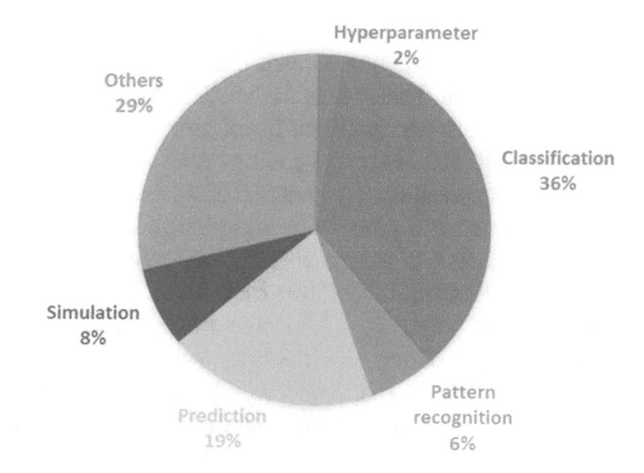

Fig. 2.4 Research domains related to machine learning algorithms (Web of science database was used; keyword: machine learning; the number of publications: 84,442)

acquisition function is to obtain an unbiased estimation from samples. In other words, acquisition function balances the number of samples from unknown (exploration) and known areas (exploitation). There are many ways to construct acquisition function, such as expected improvement and Gaussian process upper confidence bound (Snoek et al. 2012; Shahriari et al. 2015).

Although most existing hyperparameter optimization approaches were developed or investigated from computer science and engineering research areas (Zheng et al. 2019), the discussion of hyperparameter optimization is not sufficient in machine learning community (Fig. 2.4). Also, existing hyperparameter optimization approaches have a set of limitations, for example, the inappropriate assumption (i.e., hyperparameters are independent) and high demand of computing (Zheng et al. 2019).

2.3 Cyberinfrastructure and High-Performance and Parallel Computing

Cyberinfrastructure (CI) includes computing systems, data, services, tools, and virtual organizations for solving scientific problems (NSF 2007), which provides solid support for addressing computing challenge within the context in GIScience and other disciplines (Armstrong 2000; NSF 2007; Wang 2010; Yang et al. 2010; Tang et al. 2018). High-performance and parallel computing (HPC) is a part of CI, and there are three types of HPC: cluster computing, grid computing, and cloud computing (Yang et al. 2010). As mentioned by Hey et al. (2009), HPC also promotes the development of data-driven science. For GIScience field, it is necessary and meaningful to use cyberinfrastructure and HPC for exploring and understanding complex and nonlinear geographical phenomena.

There are four categories of parallel computer architectures (Wilkinson and Allen 1999): Single Instruction Stream, Single Data Stream (SISD); Single Instruction Stream, Multiple Data Stream (SIMD); Multiple Instruction Stream, Single Data Stream (MISD); and Multiple Instruction stream, Multiple Data stream (MIMD). SIMD and MIMD are two popular parallel computer architecture categories, which have many applications (Armstrong 2020). SIMD uses a single operation instruction to execute multiple processors using multiple datasets. In contrast, MIMD adopts multiple processors to run different operation executions using multiple datasets (Ding and Densham 1996; Armstrong 2020). Based on SIMD and MIMD computers, there are two popular parallel computing paradigms: multi-core and many-core. Multi-core computing is an extension of single-core computing, which includes multiple processors to execute an execution instruction(s). Typically, these multiple processors are connected by shared memory modules. Since early 2000, many researchers suggested the use of graphic processing units (GPUs; an example of many-core computing) to further accelerate computing performance for a range of studies (Tang and Bennett 2009; Nickolls and Dally 2010; Krieder et al. 2012; Zhang et al. 2015).

Based on multi-core and many-core computing paradigms, embarrassingly parallel, shared memory, and message passing are three parallel approaches (Wilkinson and Allen 1999). For the embarrassingly parallel approach, communication among processors is not necessary. In other words, tasks with no communication requirements can adopt embarrassingly parallel approach. Otherwise, shared memory and message passing can be considered. In shared memory approach, different processors access and exchange data through shared memory space. In contrast, the message passing approach does not rely on shared memory space, and each processor has its own memory space. The communication among processors for data exchange is handled via message sending or receiving. However, communication among processors can lead to a significantly reduced in computing performance. Further, decomposition is a necessary step for all parallel computing tasks, which introduces load balancing[1] issues (Wilkinson and Allen 1999).

Besides adopting suitable parallel approaches, decomposition is also an important strategy for HPC. Complete decomposition, domain decomposition, and control decomposition are three categories of decomposition methods (Wilkinson and Allen 1999). The complete decomposition method is the simplest decomposition method that divides a task into multiple independent sub-tasks. The communication among sub-tasks is not necessary. The basic idea of domain decomposition is to divide a global domain into many subdomains. However, the difference between complete decomposition and domain decomposition is that domain decomposition can apply to dependent sub-tasks with necessary communication. Based on this feature, domain decomposition is a popular approach for spatial problems. Ding and Densham (1996) summarized the scopes of applications for domain decomposition, including static data structure problems, dynamic data structure problems, and fixed domain but

[1] Load balancing is spreading tasks across processors that lead to maximize the use of computing resources, minimize response time, and avoid overloading (Wilkinson and Allen 1999).

dynamic computation problems. Control decomposition covers two decomposition methods: functional decomposition and manager/worker decomposition. Functional decomposition breaks the execution instruction into a set of sub-instructions and then allocates to multiple processors. For manager/worker decomposition, the function of the manager is to assign tasks to workers, and workers return the finished jobs to the manager. A benefit of manager/worker decomposition is efficiency because the manager dynamically distributes tasks to workers instead of allocating tasks in advance.

The advancements of cyberinfrastructure and HPC have attracted much attention in the GIScience community. From spatial statistics, Armstrong et al. (1994) developed a method for addressing the computational intensity issue when processing measures of spatial association (G). A parallel implementation for computing $G^*(d)$ statistic using grid computing was proposed by Wang et al. (2008). Hohl et al. (2016) adopted HPC to accelerate the process of space–time kernel density estimation (STKDE). In spatial simulation, a parallel agent-based model for simulating large-scale land-use opinions was presented by Tang et al. (2011). Gong et al. (2013) developed a parallel approach to handle the computational complexity of a spatial interaction simulation model. In addition, there are a number of applications of cyberinfrastructure and HPC in geovisualization. For example, Soroknine (2007) discussed a parallel implementation of visualizing big spatial data using GRASS GIS software. Tang (2013) adopted GPUs to accelerate the development of a type of area cartograms (i.e., circular cartograms).

Hyperparameter optimization typically comes with data- and computational-intensity issues. HPC has the ability to handle these issues. The use of HPC in addressing data- and computational-intensity have been discussed in existing hyperparameter optimization approaches. The adoption of computing clusters (GPU- or CPU-based) is at the center of current studies (e.g., Bergstra et al. 2013, 2015; Lorenzo et al. 2017; Falkner et al. 2018).

2.4 Evolutionary Algorithms

Evolutionary computation is a population-based machine learning algorithm, which is inspired by biological evolution (Darwinian principles of natural selection). The algorithms involved in evolutionary computation include four main types: genetic algorithms (GA), genetic programming (GP), evolutionary strategies (ES), and evolutionary programming (EP) (Eiben and Smith 2003). The ability of EAs for handling real-world problems has been discussed by several studies, see for example, Srinivas and Deb (1994), Zitzler and Thiele (1998), Deb (2001), and Xiao et al. (2002).

There are six necessary steps of EAs: representation, initialization, selection, evaluation, recombination, and mutation (Fig. 2.5). Representation is the first step of EAs. The major goal of this step is to transform real-world problems to EA-world problems using encoding methods. The context within problems from real-world is called phenotypes. Genotypes are the encoded context from real-world problems,

Fig. 2.5 The general workflow of evolutionary algorithms

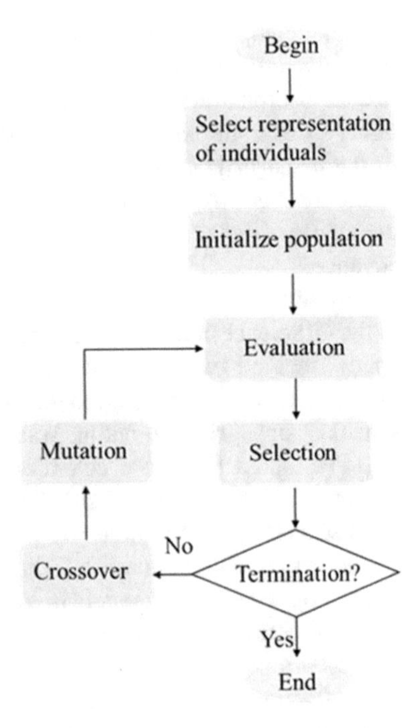

which refer to chromosomes within EAs. There are a number of available encoding methods, such as binary string, a string of integer or float numbers, and tree-based (Eiben and Smith 2003). Binary string method is the popular encoding method in EAs (Cao et al. 2014), but some studies suggest that other representations may be more effective than binary string for specific cases (Armstrong et al. 2003; Cao et al. 2014). Earlier works tried to include all hyperparameters into the binary string, but it resulted in poor scalability (Kitano 1994). Thus, how many hyperparameters in a chromosome and how to encode them post a challenge. The initialization step provides an initial set of solutions, which has the similar function as sampling.

The role of selection and evaluation is to show directions of improvement. Typically, each individual has a fitness value associated with it. The fitness values are derived from measurement indexes determined by objective functions. The selection process identifies better solutions to survive to the next generation based on fitness values. That is, solutions with higher fitness values have a higher chance of surviving, whereas lower fitness has a lower chance of surviving (Branke 1995; Xiao et al. 2002; Cao et al. 2014). However, there is still a chance that solutions with high values may perish, and solutions with low values may survive. Thus, fitness proportionate selection or elitism selection is needed for improving the generation process. Fitness proportionate selection (aka, roulette-style selection algorithm Lipowski and Lipowska 2012), is to select a part of solutions for recombination using a probability of being selected. The remaining solutions are copied into next generation without modification (Goldberg 1989). Elitism selection is to select the best solutions (e.g., top 10%) in the current generation and duplicating those best solutions directly to the next generation without modification, and then the remaining solutions are produced through variation steps. Elitism selection avoids missing the best solutions found in the previous generations, whereas roulette-style selection maintains genetic diversity.

Variation operators have two types of operators, recombination (crossover) and mutation (Eiben and Smith 2003). The role of variation operators is to create new individuals from old generations, and to maintain genetic diversity. Variation operators are accomplished by creating an offspring from two parents and the offspring contains genotypes from both sides of parents. Meanwhile, variation operators can speed up the local fine-tuning because only those survived individuals in the current generation have the chance to survive in the next generation.

Recombination (or crossover) is that the offspring chooses and exchanges one of some parts of chromosomes from parents with a certain probability. The general recombination operators include one-point, n-point, uniform, discrete, intermediate or arithmetic, and subtree (Eiben and Smith 2003). The one-point recombination is illustrated in Fig. 2.6. Besides the types of recombination operator, there exists another hyperparameter in recombination operator—crossover probability (or crossover rate). Crossover probability is used to determine how often individuals will be performed crossover operator. A larger value of crossover rate, the speed of generating offspring is fast than the smaller values. For example, if crossover probability is 60%, then the offspring has 60% chance of being created by crossover operator.

The primary function of mutation is to maintain genetic diversity and enhance the local searching capability of EAs that avoids the local minima issue. Such diversity is important for the successful application of EAs to a multi-objective optimization problem because it is related to the exploration of the search space (Deb 2001; Xiao et al. 2007). Like the recombination operator, there are a number of types of

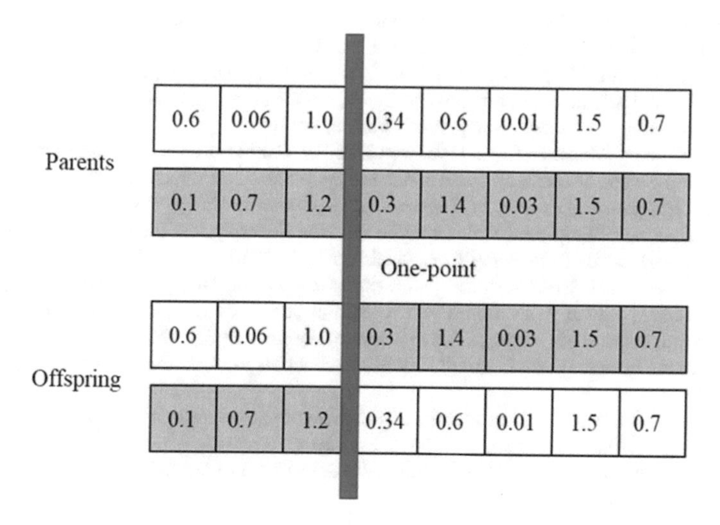

Fig. 2.6 One-point recombination of real-valued representation

mutation operator, such as bit-flip, random resetting, and Gaussian (Eiben and Smit 2011). Mutation probability (or mutation rate) is used to determine how often do parts of a chromosome mutated. For example, if mutation probability is 1%, then parts of a chromosome of an individual has 1% chance of being mutated. Typically, the value of mutation probability is small. If the probability is high (e.g., 60%), then EAs will in fact change to random search.

References

Almeida, C.M., J.M. Gleriani, Emiliano Ferreira Castejon, and B.S. Soares-Filho. 2008. Using neural networks and cellular automata for modelling intra-urban land-use dynamics. *International Journal of Geographical Information Science* 22 (9): 943–963.

Armstrong, Marc P. 2000. Geography and computational science.

Armstrong, Marc P. 2020. High performance computing for geospatial applications: A retrospective view. In *High Performance Computing for Geospatial Applications*, 9–25. Springer.

Armstrong, Marc P, Claire E Pavlik, Richard Marciano. 1994. Parallel processing of spatial statistics. *Computers and Geosciences* 20 (2): 91–104.

Armstrong, Marc P., Ningchuan Xiao, and David A. Bennett. 2003. Using genetic algorithms to create multicriteria class intervals for choropleth maps. *Annals of the Association of American Geographers* 93 (3): 595–623.

Atkinson, Peter M., and A.R.L. Tatnall. 1997. Introduction neural networks in remote sensing. *International Journal of Remote Sensing* 18 (4): 699–709.

Bergstra, James, Brent Komer, Chris Eliasmith, Dan Yamins, and David D Cox. 2015. Hyperopt: a python library for model selection and hyperparameter optimization. *Computational Science & Discovery* 8 (1): 014008.

Bergstra, James, Dan Yamins, and David D Cox. 2013. Hyperopt: A python library for optimizing the hyperparameters of machine learning algorithms. In *Proceedings of the 12th python in science conference.*

Bishop, Christopher M. 2006. *Pattern recognition and machine learning*. Springer.

Branke, Jürgen. 1995. Evolutionary algorithms for neural network design and training.

Brondino, Nair Cristina Margarido, and A.N.R. da Silva. 1999. Combining artificial neural networks and GIS for land valuation purposes. In *Proceedings of 6th international conference on computers in urban planning and urban management*, Venice, Italy.

Cao, Kai, Bo Huang, Manchun Li, and Wenwen Li. 2014. Calibrating a cellular automata model for understanding rural–urban land conversion: A Pareto front-based multi-objective optimization approach. *International Journal of Geographical Information Science* 28 (5): 1028–1046.

Chen, Sheng, S.A. Billings, and P.M. Grant. 1990. Non-linear system identification using neural networks. *International Journal of Control* 51 (6): 1191–1214.

Deb, Kalyanmoy. 2001. *Multi objective optimization using evolutionary algorithms*. Wiley.

Ding, Yuemin, and Paul J. Densham. 1996. Spatial strategies for parallel spatial modelling. *International Journal of Geographical Information Systems* 10 (6): 669–698.

Eiben, Agoston E, and James E Smith. 2003. *Introduction to evolutionary computing*, vol. 53. Springer.

Eiben, Agoston Endre, and Selmar K Smit. 2011. Evolutionary algorithm parameters and methods to tune them. In *Autonomous search*, 15–36. Springer.

Falkner, Stefan, Aaron Klein, and Frank Hutter. 2018. BOHB: Robust and efficient hyperparameter optimization at scale.

Goldberg, David E. 1989. *Genetic algorithms in search, optimization, and machine learning*. Reading, MA: Addison Wesley. Summary the applications of ga-genetic algorithm for dealing with some optimal calculations in economics.

Gong, Zhaoya, Wenwu Tang, David A. Bennett, and Jean-Claude. Thill. 2013. Parallel agent-based simulation of individual-level spatial interactions within a multicore computing environment. *International Journal of Geographical Information Science* 27 (6): 1152–1170.

Gopal, Sucharita. 2017. Artificial neural networks in geospatial analysis. *The International Encyclopedia of Geography*.

Guan, Qingfeng, Liming Wang, and Keith C. Clarke. 2005. An artificial-neural-network-based, constrained CA model for simulating urban growth. *Cartography and Geographic Information Science* 32 (4): 369–380.

Gulliford, Sarah L, Steve Webb, Carl G Rowbottom, David W Corne, David P. Dearnaley. 2004. Use of artificial neural networks to predict biological outcomes for patients receiving radical radiotherapy of the prostate. *Radiotherapy and oncology* 71 (1): 3–12

Hahnloser, Richard HR, Rahul Sarpeshkar, Misha A Mahowald, Rodney J Douglas, and H Sebastian Seung. 2000. Digital selection and analogue amplification coexist in a cortex-inspired silicon circuit. Nature 405 (6789): 947–951

Hey, Anthony JG, Stewart Tansley, and Kristin Michele Tolle. 2009. *The fourth paradigm: data-intensive scientific discovery*, vol. 1. Redmond, WA: Microsoft research.

Hohl, Alexander, Eric Delmelle, Wenwu Tang, and Irene Casas. 2016. Accelerating the discovery of space-time patterns of infectious diseases using parallel computing. *Spatial and Spatio-Temporal Epidemiology* 19: 10–20.

Hornik, Kurt, Maxwell Stinchcombe, and Halbert White. 1989. Multilayer feedforward networks are universal approximators. *Neural Networks* 2 (5): 359–366.

Karsoliya, Saurabh. 2012. Approximating number of hidden layer neurons in multiple hidden layer BPNN architecture. *International Journal of Engineering Trends and Technology* 3 (6): 714–717.

Kitano, Hiroaki. 1994. Neurogenetic learning: An integrated method of designing and training neural networks using genetic algorithms. *Physica D: Nonlinear Phenomena* 75 (1–3): 225–238.

Krieder, Scott, Ben Grimmer, and Ioan Raicu. 2012. Early experiences in running many-task computing workloads on gpgpus. *XSEDE Poster Session*.

Li, Xia, and Anthony Gar-On Yeh. 2002. Neural-network-based cellular automata for simulating multiple land use changes using GIS. *International Journal of Geographical Information Science* 16 (4):323–343.

Limsombunchai, Visit. 2004. House price prediction: hedonic price model versus artificial neural network. New Zealand Agricultural and Resource Economics Society Conference.

Lipowski, Adam, and Dorota Lipowska. 2012. Roulette-wheel selection via stochastic acceptance. *Physica A: Statistical Mechanics and its Applications* 391 (6): 2193–2196.

Lorenzo, Pablo Ribalta, Jakub Nalepa, Luciano Sanchez Ramos, and José Ranilla Pastor. 2017. Hyper-parameter selection in deep neural networks using parallel particle swarm optimization. In *Proceedings of the Genetic and Evolutionary Computation Conference Companion*.

Mas, Jean F., and Juan J. Flores. 2008. The application of artificial neural networks to the analysis of remotely sensed data. *International Journal of Remote Sensing* 29 (3): 617–663.

Mellit, Adel, and Soteris A Kalogirou. 2008. Artificial intelligence techniques for photovoltaic applications: A review. *Progress in Energy and Combustion Science* 34 (5): 574–632.

Močkus, J. 1975. On Bayesian methods for seeking the extremum. Optimization Techniques IFIP Technical Conference.

Nasrabadi, Nasser M. 2007. Pattern recognition and machine learning. *Journal of Electronic Imaging* 16 (4): 049901.

Neaupane, Krishna Murari, and NR Adhikari. 2006. Prediction of tunneling-induced ground movement with the multi-layer perceptron. *Tunnelling and Underground Space Technology* 21 (2): 151–159.

Nickolls, John, and William J. Dally. 2010. The GPU computing era. *IEEE Micro* 30 (2): 56–69.

Niu, Jiqiang, Wenwu Tang, Feng Xu, Xiaoyan Zhou, and Yanan Song. 2016. Global research on artificial intelligence from 1990–2014: Spatially-explicit bibliometric analysis. *ISPRS International Journal of Geo-Information* 5 (5): 66.

NSF. 2007. Cyberinfrastructure vision for 21st century discovery. https://www.nsf.gov/pubs/2007/nsf0728/.

Omrani, Hichem, Amin Tayyebi, and Bryan Pijanowski. 2017. Integrating the multi-label land-use concept and cellular automata with the artificial neural network-based Land Transformation Model: An integrated ML-CA-LTM modeling framework. *Giscience and Remote Sensing* 54 (3): 283–304.

Openshaw, Stan, and Christine Openshaw. 1997. *Artificial intelligence in geography*: Wiley.

Pijanowski, Bryan C., Snehal Pithadia, Bradley A. Shellito, and Konstantinos Alexandridis. 2005. Calibrating a neural network-based urban change model for two metropolitan areas of the Upper Midwest of the United States. *International Journal of Geographical Information Science* 19 (2): 197–215.

Pijanowski, Bryan C, Amin Tayyebi, Jarrod Doucette, Burak K Pekin, David Braun, and James Plourde. 2014. A big data urban growth simulation at a national scale: Configuring the GIS and neural network based Land Transformation Model to run in a High Performance Computing (HPC) environment. *Environmental Modelling & Software* 51: 250–268.

Roberts, Craig A, and Nii O Attoh-Okine. 1998. A comparative analysis of two artificial neural networks using pavement performance prediction. *Computer-Aided Civil and Infrastructure Engineering* 13 (5): 339–348.

Rosenblatt, Frank. 1958. The perceptron: A probabilistic model for information storage and organization in the brain. *Psychological Review* 65 (6): 386.

Sanger, Terence D. 1989. Optimal unsupervised learning in a single-layer linear feedforward neural network. *Neural Networks* 2 (6): 459–473.

Shahriari, Bobak, Kevin Swersky, Ziyu Wang, Ryan P Adams, and Nando De Freitas. 2015. Taking the human out of the loop: A review of Bayesian optimization. *Proceedings of the IEEE* 104 (1): 148–175.

Snoek, Jasper, Hugo Larochelle, and Ryan P Adams. 2012. Practical bayesian optimization of machine learning algorithms. *Advances in Neural Information Processing Systems.*

Sorokine, Alexandre. 2007. Implementation of a parallel high-performance visualization technique in GRASS GIS. *Computers and Geosciences* 33 (5): 685–695.

Specht, Donald F. 1990. Probabilistic neural networks. Neural Networks 3 (1): 109–118.

Srinivas, Nidamarthi, and Kalyanmoy Deb. 1994. Muiltiobjective optimization using nondominated sorting in genetic algorithms. *Evolutionary Computation* 2 (3): 221–248.

Stathakis, D. 2009. How many hidden layers and nodes? *International Journal of Remote Sensing* 30 (8): 2133–2147.

Tang, Wenwu. 2013. Parallel construction of large circular cartograms using graphics processing units. *International Journal of Geographical Information Science* 27 (11): 2182–2206.

Tang, Wenwu, and David A Bennett. 2009. Parallel agent-based modelling of land-use opinion dynamics using graphics processing units. In *Proceedings of the 10th International Conference on GeoComputation.*

Tang, Wenwu, David A. Bennett, and Shaowen Wang. 2011. A parallel agent-based model of land use opinions. *Journal of Land Use Science* 6 (2–3): 121–135.

Tang, Wenwu, Wenpeng Feng, Jing Deng, Meijuan Jia, and Huifang Zuo. 2018. Parallel Computing for Geocomputational Modeling. In *GeoComputational Analysis and Modeling of Regional Systems*, 37–54. Springer.

Wanas, Nayaer, Gasser Auda, Mohammad S Kamel, and FAKF Karray. 1998. On the optimal number of hidden nodes in a neural network. In *IEEE Canadian Conference on Electrical and Computer Engineering.*

Wang, Shaowen. 2010. A CyberGIS framework for the synthesis of cyberinfrastructure, GIS, and spatial analysis. *Annals of the Association of American Geographers* 100 (3): 535–557.

Wang, Shaowen, Mary Kathryn Cowles, and Marc P. Armstrong. 2008. Grid computing of spatial statistics: Using the TeraGrid for G (d) analysis. *Concurrency and Computation: Practice and Experience* 20 (14): 1697–1720.

Wilkinson, Barry, and Michael Allen. 1999. *Parallel programming*, vol. 999. Prentice hall Upper Saddle River, NJ.

Williams, Christopher KI, and Carl Edward Rasmussen. 2006. *Gaussian processes for machine learning*, vol. 2. MIT press Cambridge, MA.

Xiao, Ningchuan, David A. Bennett, and Marc P. Armstrong. 2002. Using evolutionary algorithms to generate alternatives for multiobjective site-search problems. *Environment and Planning A* 34 (4): 639–656.

Xiao, Ningchuan, David A. Bennett, and Marc P. Armstrong. 2007. Interactive evolutionary approaches to multiobjective spatial decision making: A synthetic review. *Computers, Environment and Urban Systems* 31 (3): 232–252.

Xu, Xiaodong, Wei Li, Qiong Ran, Qian Du, Lianru Gao, and Bing Zhang. 2017. Multisource remote sensing data classification based on convolutional neural network. *IEEE Transactions on Geoscience and Remote Sensing* 56 (2): 937–949.

Yang, Chaowei, Robert Raskin, Michael Goodchild, and Mark Gahegan. 2010. Geospatial cyberinfrastructure: Past, present and future. *Computers, Environment and Urban Systems* 34 (4): 264–277.

Yeh, Anthony Gar-On, and Xia Li. 2003. Simulation of development alternatives using neural networks, cellular automata, and GIS for urban planning. *Photogrammetric Engineering Remote Sensing* 69 (9): 1043–1052.

Zhang, Lingqi, Tianyi Wang, Zhenyu Jiang, Qian Kemao, Yiping Liu, Zejia Liu, Liqun Tang, and Shoubin Dong. 2015. High accuracy digital image correlation powered by GPU-based parallel computing. *Optics Lasers in Engineering* 69: 7–12.

Zheng, Minrui, Wenwu Tang, and Xiang Zhao. 2019. Hyperparameter optimization of neural network-driven spatial models accelerated using cyber-enabled high-performance computing. *International Journal of Geographical Information Science*. https://doi.org/10.1080/13658816.2018.1530355.

Zitzler, Eckart, and Lothar Thiele. 1998. Multiobjective optimization using evolutionary algorithms—a comparative case study. *International conference on parallel problem solving from nature*.

Chapter 3
Methodology

3.1 Overview

In previous sections, I identified current hyperparameter optimization approaches have a number of limitations, such as inappropriate assumption and high computing cost. In this book, I propose a spatially explicit hyperparameter optimization framework that focuses on addressing data and computational-intensity challenges, considering spatial dependence, and exploring local landscape of search space of hyperparameters.

The generic framework of spatially explicit hyperparameter optimization is illustrated in Fig. 3.1. Three modules are covered in spatially explicit hyperparameter optimization: generation of sampled hyperparameters, evaluation of sampled hyperparameters, and hyperparameter analysis. Generation of sampled hyperparameters determines the search space of hyperparameters (i.e., hyperparameters that need to be optimized) and how to acquire sampled hyperparameters. Evaluation of sampled hyperparameters module examines the performance of sampled hyperparameters through three aspects: whether the sample size is enough; whether the number of repetitions is enough; and whether results from samples are robust. Generalization performance of the search space, computing performance, and model performance are evaluated in the hyperparameter analysis module. For detailed explanation see Chap. 4 and Zheng et al.'s study (2019).

Further, the framework comprises three components: automatic search of hyperparameters, spatial prediction of hyperparameter space, and acceleration of hyperparameter search. Each component contributes to one or more modules. The major contribution of automatic search of hyperparameters component is to automate the process of spatially explicit hyperparameter optimization. Acceleration of hyperparameter search component is responsible for handling data- and computational-intensity issues of spatially explicit hyperparameter optimization. Moreover, spatial prediction of hyperparameter space component links to evaluation of sampled hyperparameters and hyperparameter analysis modules. I will discuss the core parts of these three components in the following sections. That is, Sects. 3.2, 3.3 and 3.4

M. Zheng, *Spatially Explicit Hyperparameter Optimization for Neural Networks*,
https://doi.org/10.1007/978-981-16-5399-5_3

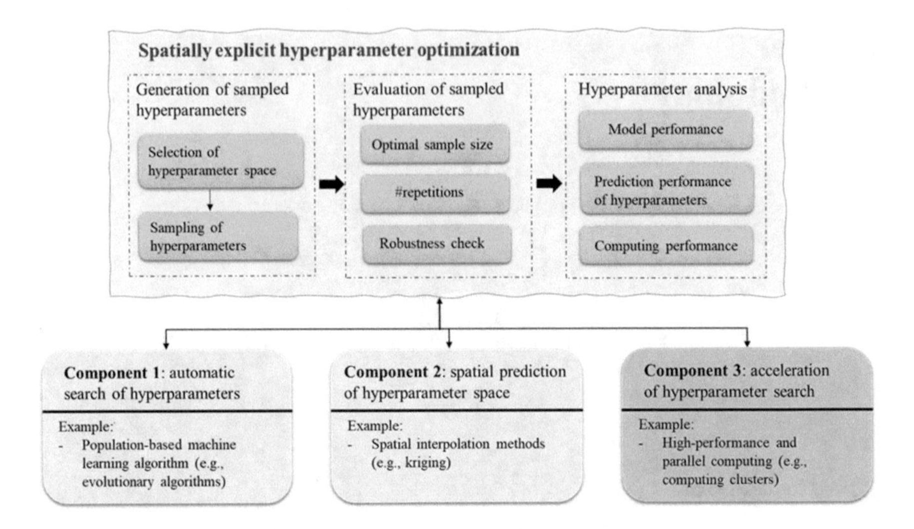

Fig. 3.1 Generic framework of spatially explicit hyperparameter optimization

will cover automatic search of hyperparameters (component 1), spatial prediction of hyperparameter space (component 2), and acceleration of hyperparameter search (component 3), respectively.

Spatially explicit hyperparameter optimization that is proposed in this book covers three research communities—spatial modeling, cyberinfrastructure, and machine learning algorithms (Fig. 3.2). The major contributions of this book are: (1) considering spatial dependence between hyperparameters, (2) exploring and visualizing the landscape of the search space of hyperparameters, and (3) improving the computing performance at both model- and computing-levels.

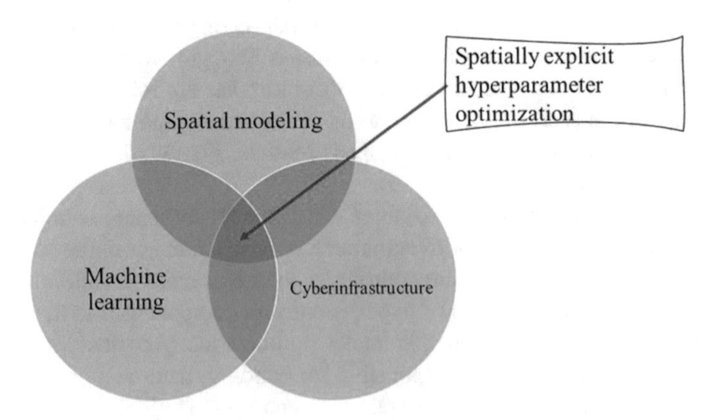

Fig. 3.2 The Venn diagram of this book

3.2 Component 1—Automatic Search of Hyperparameters

Machine learning algorithms have great potential capabilities for discovering of the relationships from data through structural and functional aspects (Mitchell 1997). Several geographers have placed emphasis on the use of machine learning algorithms to spatial modeling (Batty et al. 1999; Sui and Maggio 1999; Pijanowski et al. 2002, 2014; Li and Yeh 2002). However, the discussion of how to select hyperparameters for machine learning algorithm-based spatial modeling is inadequate in GIScience community.

Since most machine learning algorithms contain more than one hyperparameters, there is a challenge for finding multiple hyperparameters at the same time. There has been a great interest in evolving neural networks through population-based hyperparameter optimization since 1990s (Belew et al. 1990; Yao 1994; Zhang et al. 2000; Yu et al. 2008). As one of the global optimization methods, EAs derive their behavior from a metaphor of the processes of evolution in nature. EA-based hyperparameter optimization method has a series of advantages, for example, (1) EAs can increase the efficiency and robustness of the optimum seeking process, (2) EAs are suitable for complex evaluation functions, and (3) EAs can reduce the occurrence of local optimum (Ding et al. 2013). Although EA-based hyperparameter optimization approaches have been successfully implemented in the computer science field, the discussion of using EA-based hyperparameter optimization in GIScience is insufficient. Meanwhile, existing EA-based hyperparameter optimization approaches have some limitations, which are discussed in the previous section (Sect. 2.4). Further, the methods from GIScience may benefit current EA-based hyperparameter optimization, which may further improve the computing performance.

A number of studies have been shown that incorporating prior knowledge can significantly improve the learning process (Schwarz and Ocenasek 2000; Pitiot et al. 2009). Much recent research has been devoted to learning algorithms for architectures of machine learning algorithms, such as the deep brief network (Erhan et al. 2010). They found that supervised learning tasks with an unsupervised learning component can improve model performance. The results from unsupervised learning component can be seen as prior knowledge. In EAs, the performance of a search process is dependent on the topology of the fitness landscape (Sareni and Krahenbuhl 1998; Ratle 2001; Casas 2015). The approach proposed in this book acquires the topology of the fitness landscape from a series of prior knowledge based on the features of search space.

The main idea of this component is to build a model based on fitness values from EAs, and using this model as a guideline to do next search steps. A flowchart of the working principle of the EA-based hyperparameter optimization is shown in Fig. 3.3. The novelty of this approach has two aspects: firstly, it allows the incorporation of spatial features into the EA. A spatial statistical model will be developed based on the regular fitness function ($fitness_R$; which is calculated without statistical model), this

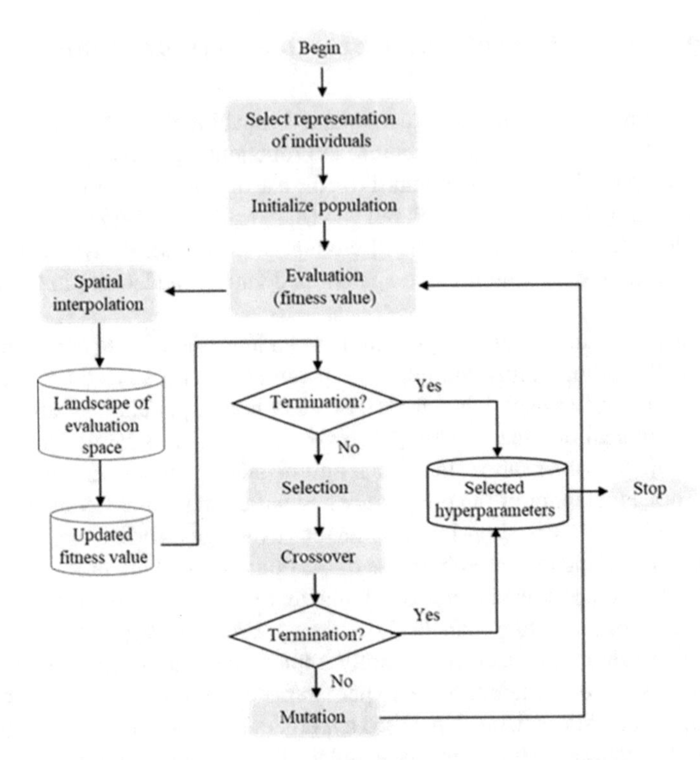

Fig. 3.3 A flowchart of the working principle of an automated EA-based spatially explicit hyperparameter optimization

step is named as knowledge-based fitness function ($fitness_K$). Since the knowledge-based fitness function is calculated from the regular fitness function, the knowledge-based fitness value will be updated for each generation. The final fitness function ($fitness_F$) involves knowledge-based fitness function and regular fitness function. The final fitness function is computed as follows:

$$fitness_F = w_R fitness_R + w_K fitness_K \tag{3.1}$$

where w_R and w_K are the weights of regular fitness function and knowledge-based fitness function, the sum of weights is equal to 1. Secondly, maps of the evaluation landscape will generate for each generation. These landscape maps could show the path of the search process of hyperparameter optimization. This model is used to generate the generalization performance of hyperparameters.

3.3 Component 2—Spatial Prediction of Hyperparameter Space

Sampling is the most commonly used method in current hyperparameter optimization. However, those sampled hyperparameters cannot represent the entire hyperparameter space. Thus, a method that can convert discrete points to continuous space is needed. Spatial interpolation, which is a component of spatial statistics, provides such an ability that transforms discrete points into continuous space with consideration of spatial dependence.

Spatial interpolation has been applied in different disciplines, such as geography and environmental sciences (Li and Heap 2014). Spatial interpolation methods can be divided into two categories: point interpolation and areal interpolation. Point interpolation focuses on point data, such as daily temperature, whereas areal interpolation deals with the data that aggregated over areas, such as average pollution levels for counties (Lam 1983). A number of spatial interpolation methods have been developed. Lam (1983) divided these spatial interpolation methods into four types: exact and approximate methods for point interpolation, and non-volume-preserving (the same as the exact and approximate for point interpolation) and volume-preserving methods for areal interpolation. Examples of exact methods include weighting, kriging, and splines. In contrast, Fourier series, least-square fitting with splines, and distance-weighted least-squares are examples of approximate methods (Lam 1983). Meanwhile, Li and Heap (2014) summarized interpolation methods into three categories: non-geostatistical methods (e.g., nearest neighbors and inverse distance weighting), geostatistical methods (discuss later), and combined methods (e.g., trend surface analysis combined with kriging).

Kriging is a key component of spatial interpolation methods (Krige 1978). It is a linear interpolation method, and it uses a semivariogram model to investigate spatial autocorrelation (spatial dependence). The semivariogram model is used for measuring the spatial dependence between sample points. The general assumption for kriging is that the semivariogram is known. Moreover, semivariogram has three characteristics: range, sill, and nugget (Matheron 1963). Besides those, kriging is called spatial BLUE (best linear unbiased prediction) (Le and Zidek 2006). There are two groups of kriging: univariate (one-dimension) and multivariate (multi-dimension). Ordinary kriging (OK) and universal kriging (UK) are examples of univariate interpolation methods, and cokriging is the example of multivariate interpolation methods (Wackernagel et al. 2002; Li and Heap 2014). Besides linear interpolation methods, there are a number of non-linear and machine learning-based interpolation methods, for example, support vector machines (Li et al. 2011) and neural networks (Klesk 2008).

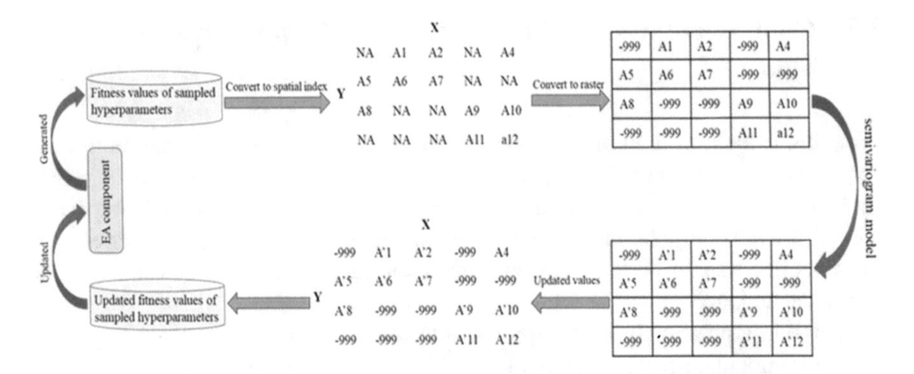

Fig. 3.4 The flowchart of spatial prediction using two hyperparameters (−999 is null value in the raster dataset; A stands for original value; A' is updated value through semivariogram model)

Although there exist a number of spatial interpolation methods, kriging is the most commonly used interpolation method in GIScience field, particularly for the OK and UK. OK is derived from an assumption of intrinsic stationarity (the mean is constant). The variation of kriging that can handle trends is called UK. In other words, the mean function of UK is not constant. In this book, I examine the capabilities of spatial interpolation in spatially explicit hyperparameter optimization. Thus, UK is adopted to predict the continuous patterns from selected hyperparameters.

The major objective of this component is to build a semivariogram model of the fitness landscape from the generations of EA. Figure 3.4 shows the flowchart of the spatial prediction component. First, we collect all sampled hyperparameter sets (i.e., chromosomes of EA) and their related fitness values. Second, we convert sampled hyperparameter sets into the spatial index. For example, we have two hyperparameters (A and B), those two hyperparameters can be seen as longitude (A) and latitude (B) in a spatial dimension. Then, the result is transformed into a grid format with attributes assigned to every cell, indicating the fitness values. And then, we build a semivariogram model based on these fitness values. After building a semivariogram, we re-calculate those fitness values and return these updated values to EA. Furthermore, we create a final map of the fitness landscape using all sampled hyperparameter sets from the entire EA process. This final map is used to show the prediction of generalization performance. The detailed illustration of this component is discussed in Sect. 4.4.3, as well as Zheng et al.'s study (2019).

3.4 Component 3—Acceleration of Hyperparameter Search

While hyperparameter optimization also has data- and computational-intensity issues, those issues are still a challenge in current hyperparameter optimization studies. HPC, as a part of CI, provides powerful computing capability to handle data- and computational-intensity issues (see Sect. 2.3 for more information). In the past years, a number of applications and extensions of using HPC to address data and computational-intensity issues have been constructed in current hyperparameter optimization study. The use of computing clusters (GPU or CPU-based) is the mainstream of current studies (Bergstra et al. 2013; Bergstra et al. 2015; Lorenzo et al. 2017; Falkner et al. 2018). Young et al. (2015) used HPC platform to address the computing challenge in hyperparameter optimization. During their works, they found that their approach (EA-based hyperparameter optimization) are unable to maximize the use of available computing resources. In other words, some computing processors finish their works earlier and stay idle because the complexity of a model with different hyperparameters is different. Hence, an appropriate load balancing strategy and decomposition strategy are necessary in order to achieve a more efficient performance.

The search space of hyperparameters can be seen as an infinite space. Using brute force search to find suitable hyperparameters is often infeasible. Thus, sampling is needed in hyperparameter optimization. Each model with different hyperparameter sets is independent, and its performance has little influence on other models. Based on this feature, embarrassingly parallel computing is utilized here because this computation method is suitable for tasks that do not need to communicate with each other. In order to achieve the best performance of HPC, I decompose sampled hyperparameters into a series of subsets. More specifically, each subset contains a single combination of hyperparameters. And dynamic load balancing is used in order to address the workload imbalance issue. Computing resources can use their full potential when involving a dynamic load balancing strategy. When the status of a processor is idle, then a new task will be immediately assigned to it.

References

Batty, Michael, Yichun Xie, and Zhanli Sun. 1999. Modeling urban dynamics through GIS-based cellular automata. *Computers, Environment and Urban Systems* 23 (3): 205–233.

Belew, Richard K, John McInerney, and Nicol N Schraudolph. 1990. Evolving networks: Using the genetic algorithm with connectionist learning.

Bergstra, James, Dan Yamins, and David D Cox. 2013. Hyperopt: A python library for optimizing the hyperparameters of machine learning algorithms. In: *Proceedings of the 12th python in science conference.*

Bergstra, James, Brent Komer, Chris Eliasmith, Dan Yamins, and David D Cox. 2015. Hyperopt: A python library for model selection and hyperparameter optimization. *Computational Science & Discovery* 8 (1): 014008.

Casas, Noe. 2015. Genetic algorithms for multimodal optimization: A review. *arXiv preprint* arXiv: 1508.05342.

Ding, Shifei, Hui Li, Chunyang Su, Junzhao Yu, and Fengxiang Jin. 2013. Evolutionary artificial neural networks: A review. *Artificial Intelligence Review* 39 (3): 251–260.

Erhan, Dumitru, Yoshua Bengio, Aaron Courville, Pierre-Antoine Manzagol, Pascal Vincent, and Samy Bengio. 2010. Why does unsupervised pre-training help deep learning? *Journal of Machine Learning Research* 11 : 625–660.

Falkner, Stefan, Aaron Klein, and Frank Hutter. 2018. BOHB: Robust and efficient hyperparameter optimization at scale. *arXiv preprint arXiv*.

Klesk, Przemyslaw. 2008. Construction of a neurofuzzy network capable of extrapolating (and interpolating) with respect to the convex hull of a set of input samples in Rn. *IEEE Transactions on Fuzzy Systems* 16 (5): 1161–1179.

Krige, Danie G. 1978. *Lognormal-de Wijsian geostatistics for ore evaluation*. South African Institute of mining and metallurgy Johannesburg.

Lam, Nina Siu-Ngan. 1983. Spatial interpolation methods: A review. *The American Cartographer* 10 (2): 129–150.

Le, Nhu D, and James V Zidek. 2006. *Statistical analysis of environmental space-time processes*. Springer Science & Business Media.

Li, Xia, and Anthony Gar-On Yeh. 2002. Neural-network-based cellular automata for simulating multiple land use changes using GIS. *International Journal of Geographical Information Science* 16 (4): 323–343.

Li, Jin, Andrew D Heap, Anna Potter, James J, and Daniell. 2011. Application of machine learning methods to spatial interpolation of environmental variables. *Environmental Modelling and Software* 26 (12): 1647–1659.

Li, Jin, and Andrew D. Heap. 2014. Spatial interpolation methods applied in the environmental sciences: A review. *Environmental Modelling & Software* 53: 173–189.

Lorenzo, Pablo Ribalta, Jakub Nalepa, Luciano Sanchez Ramos, and José Ranilla Pastor. 2017. Hyper-parameter selection in deep neural networks using parallel particle swarm optimization. In: *Proceedings of the genetic and evolutionary computation conference companion*.

Matheron, Georges. 1963. Principles of geostatistics. *Economic Geology* 58 (8): 1246–1266.

Mitchell, Tom M. 1997. Machine learning. 1997. *Burr Ridge, IL: Mcgraw Hill* 45 (37): 870–877.

Pijanowski, Bryan C., Daniel G. Brown, Bradley A. Shellito, and Gaurav A. Manik. 2002. Using neural networks and GIS to forecast land use changes: A land transformation model. *Computers, Environment and Urban Systems* 26 (6): 553–575.

Pijanowski, Bryan C, Amin Tayyebi, Jarrod Doucette, Burak K Pekin, David Braun, and James Plourde. 2014. A big data urban growth simulation at a national scale: Configuring the GIS and neural network based land transformation model to run in a high performance computing (HPC) environment. *Environmental Modelling & Software* 51: 250–268.

Pitiot, Paul, Thierry Coudert, Laurent Geneste, and Claude Baron. 2009. A priori knowledge integration in evolutionary optimization. In: *International conference on artificial evolution (evolution artificielle)*.

Ratle, Alain. 2001. Kriging as a surrogate fitness landscape in evolutionary optimization. *AI EDAM* 15 (1): 37–49.

Sareni, Bruno, and Laurent Krahenbuhl. 1998. Fitness sharing and niching methods revisited. *IEEE Transactions on Evolutionary Computation* 2 (3): 97–106.

Schwarz, Josef, and Jiři Ocenasek. 2000. A problem knowledge-based evolutionary algorithm KBOA for hypergraph bisectioning. In: *Proceedings of the 4th joint conference on knowledge-based software engineering*. IOS Press.

Sui, D.Z., and R.C. Maggio. 1999. Integrating GIS with hydrological modeling: Practices, problems, and prospects. *Computers, Environment and Urban Systems* 23 (1): 33–51.

Wackernagel, H.L., Bertino, J.P., Sierra, and J.G.D. Río. 2002. Multivariate kriging for interpolating with data from different sources. In *Quantitative methods for current environmental issues*, 57–75. Springer.

Yao, Xin. 1994. The evolution of connectionist networks. In *Artificial intelligence and creativity*, 233–243. Springer.

Young, Steven R, Derek C Rose, Thomas P Karnowski, Seung-Hwan Lim, and Robert M Patton. 2015. Optimizing deep learning hyper-parameters through an evolutionary algorithm. In: *Proceedings of the workshop on machine learning in high-Performance computing environments.*

Yu, Jianbo, Shijin Wang, and Lifeng Xi. 2008. Evolving artificial neural networks using an improved PSO and DPSO. *Neurocomputing* 71 (4–6): 1054–1060.

Zhang, Chunkai, Huihe Shao, and Yu Li. 2000. Particle swarm optimisation for evolving artificial neural network. In: *Systems, man, and cybernetics, 2000 IEEE international conference on.*

Zheng, Minrui, Wenwu Tang, and Xiang Zhao. 2019. Hyperparameter optimization of neural network-driven spatial models accelerated using cyber-enabled high-performance computing. *International Journal of Geographical Information Science*. https://doi.org/10.1080/13658816.2018.1530355.

Chapter 4
Study I. Hyperparameter Optimization of Neural Network-Driven Spatial Models Accelerated Using Cyber-Enabled High-Performance Computing

In this chapter, the major purpose is to examine the feasibility and necessity of spatially explicit hyperparameter optimization and to evaluate the performance of hyperparameter optimization in neural network-based spatial models (links to objective 1).

4.1 Introduction

Geographic modeling with spatial or spatiotemporal data is an important research topic in the domain of Geographic Information Science (GIScience). Early studies were primarily concerned with the use of statistical methods to analyze spatial data and most of these models are linear. However, geospatial phenomena are often driven by nonlinear mechanisms. Linear methods may be ill-suited to the spatially explicit modeling of geospatial phenomena. Due to the need of non-linear approaches, artificial neural networks (ANNs) have been applied extensively in a variety of geography-related fields (Dai et al. 2005; Goethals et al. 2007; Mas and Flores 2008; Grekousis and Photis 2014). While there have been some studies reported to investigate the optimal configuration of neural networks (Paola and Schowengerdt 1995; Kavzoglu and Mather 2003; Goethals et al. 2007), the impact of neural network configuration on model performance has not been adequately investigated (Mas and Flores 2008). In this study, we aim to investigate the capability of hyperparameter optimization in neural network-based spatially explicit models, and to examine whether and how hyperparameter optimization can be informed by such fields as GIScience. We focus on addressing three challenges associated with hyperparameter optimization of spatially explicit modeling based on ANNs. First, the determination of neural networks configuration is fundamentally a model selection problem. Most studies

This chapter is derived from an article published in International Journal of Geographical Information Science, October 12th, 2018, copyright Taylor & Francis.

M. Zheng, *Spatially Explicit Hyperparameter Optimization for Neural Networks*,
https://doi.org/10.1007/978-981-16-5399-5_4

in the GIScience domain used a trial-and-error approach to develop the structure of ANNs because of the complicated nature of configuring ANNs (Heermann and Khazenie 1992; Li and Yeh 2002; Pijanowski et al. 2005, 2014). However, the trial-and-error approach is based on a manual mechanism and is often subjective. Hyperparameter optimization is a model selection approach that can help search for optimal neural network-based models (Bergstra and Bengio 2012). Hyperparameters are parameters of an algorithm (here ANNs) that support the determination of standard parameters of a model (spatial model here) from data. Examples of hyperparameters are learning rate and momentum for ANNs. A series of studies on hyperparameter optimization have been reported in the domain of computer science and engineering (Bergstra et al. 2011, 2013). Yet, there are no such studies that have been conducted in the domain of spatially explicit modeling. Second, hyperparameter optimization has been relatively well investigated in the fields of computer science, but, challenges remain for the study of hyperparameter optimization (Claesen and De Moor 2015). For example, existing applications of hyperparameter optimization are depended on the assumption that all potential combinations of hyperparameters are independent (as required by conventional statistics). However, the assumption of independence is often violated for practical cases and data, especially for those that contain spatial or temporal features, which has been recognized in such domains as statistics (Handcock and Wallis 1994), ecology (Tilman and Kareiva 1997), economics (Fujita et al. 1999), and geography (Tobler 1970; Anselin et al. 1996). In particular, spatial dependence in geographic space, as suggested in the First Law of Geography (Tobler 1970) has been a central component that drives the advancement of spatial statistics. For instance, spatial autocorrelation approaches have been developed to quantify and evaluate dependence in spatial data (Moran 1950; Anselin 1995). It is thus inevitable that dependence among sampled observations exists in the hyperparameter space. More importantly, without appropriately handling this form of dependence among data, analysis or modeling results may be biased (McDonald 2009). Thus, taking into account the hyperparameter dependence in the study of hyperparameter optimization is of necessity. However, no studies have been devoted to handling the hyperparameter dependence issue yet. While spatial statistics may offer insight into this issue, how to handle the dependence in the hyperparameter space using spatial statistical methods poses a challenge for the study of hyperparameter optimization. Third, the searching process of hyperparameter optimization often consumes huge amounts of computing–i.e. the hyperparameter optimization poses a computational challenge. High performance computing (HPC), driven by cyberinfrastructure (Atkins 2003), can help resolve this challenge. HPC is based on the use of a collection of computing elements (e.g. CPUs) for problem-solving in a parallel manner. In the past few decades, the use of HPC to accelerate computationally demanding hyperparameter optimization has been demonstrated by a set of applications (Bergstra et al. 2013; Kotthoff et al. 2016). However, most HPC applications of hyperparameter optimization only focus on the parallelization of the traversal of the hyperparameter space—that is, at the computing level. Leveraging characteristics of the hyperparameter space (at the method level) to facilitate the use of HPC for accelerated hyperparameter optimization (at the computing level) has been inadequately studied. Therefore, in this article, we place our focus on tackling the three challenges facing hyperparameter optimization for neural network-driven spatially explicit modeling identified above. Our specific objectives are to (1) examine the capability of hyperparameter optimization in spatially explicit modeling based on neural networks, (2) evaluate how spatial statistical methods (i.e. in the domain of GIScience) may, in turn, help the hyperparameter optimization approach (originated from the domain of computer science), and (3) explore the potential of further accelerating HPC-enabled hyperparameter optimization based on characteristics of hyperparameter space. The rest of this article is organized in the following manner. Section 4.2 provides a review of previous work in the application of ANN and hyperparameter optimization. After discussing the study area and data in Sect. 4.3, the method of automated hyperparameter optimization approach is presented in Sect. 4.4. Section 4.5 shows experimental results and discusses these results in Sect. 4.6. Section 4.7 provides concluding remarks.

4.2 Literature Review

This section presents the background of ANNs and hyperparameter optimization. These concepts include architectures and learning schemes.

4.2.1 Artificial Neural Networks

Artificial neural networks (ANNs) are an inductive machine learning approach that resembles the brain functions of human beings or animals for problem-solving (Openshaw and Openshaw 1997). ANNs are one of the powerful approaches in scientific and engineering applications when used to predict or recognize patterns. ANNs can learn from data to improve their performance and adapt themselves as more data are available. ANNs offer an alternative way of constructing complex models by allowing for representing nonlinear relationships via directly learning from data (Hornik et al. 1989; Chen et al. 1990). There are different types of ANNs regarding the network architecture: for example, feed-forward neural network, radial basis function network, and recurrent networks. Although various types of neural network models have been developed, the one that is most widely used is the feed-forward neural network. Figure 4.1 illustrates the structure of a multilayer feed-forward neural network. Typically, in a feed-forward neural network, there are three types of layers: input layer, hidden layer, and output layer. A neuron is a building block of computation, and neurons and their interconnections constitute a neural network that can be used for problem-solving. Each neuron in the input layer corresponds to an input variable. Each connection of neurons (e.g. from the input to the subsequent neuron) is associated with a weight that can be adjusted. The training of a neural network is to use learning algorithms (i.e. supervised learning algorithm) to adjust these weights

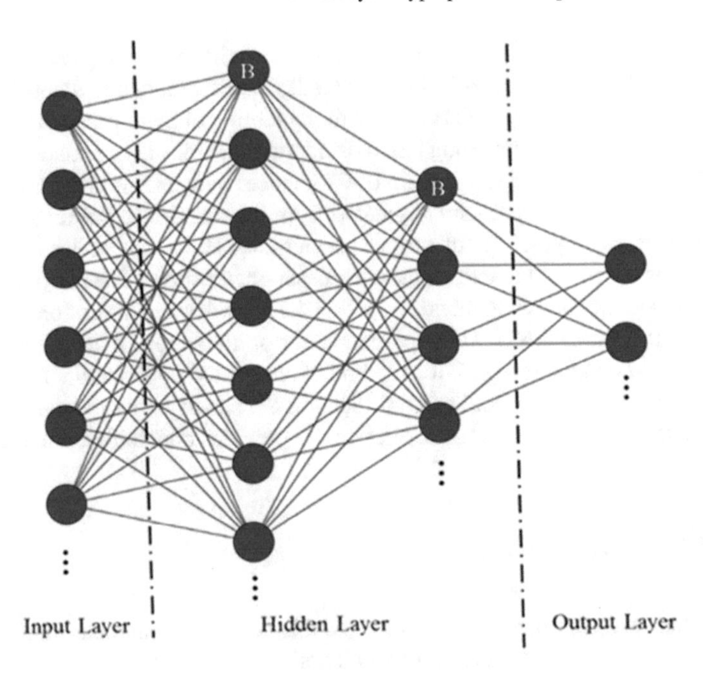

Fig. 4.1 Basic structure of a feedforward neural network (B stands for bias node; network is fully connected)

to minimize the error in outputs, thus determining the optimal network setting for problem-solving.

Supervised learning algorithm is often used to train feed-forward neural networks. Supervised learning algorithm has the capability of learning to map a set of inputs to one or more outputs by adjusting weights. A number of learning strategies have been developed, and the most popular one is the backpropagation learning algorithm introduced by Rumelhart et al. (1988). Backpropagation algorithm is based on the iterative gradient descent search mechanism. Once the weights of the network are initialized, input nodes are presented to the network and propagated forward to calculate the output value. An activation function in a neural network transforms an input or set of inputs to the output of that node. This transformation can be linear or nonlinear. Commonly used activation functions for neural networks are sigmoid (logistic), hyperbolic tangent (tanh), polynomial and linear (Specht 1990). Also, there exist other activation functions as the emergence and development of deep neural networks, for example, ReLU (rectified linear unit; see (Hahnloser et al. 2000) and Softmax (Bishop 2006)).

As one of the machine learning approaches, ANNs have a wide range of applications in such domains as pattern recognition, classification, and optimization (Maa and Schanblatt 1992; Dreiseitl and Ohno-Machado 2002; Demuth et al. 2014). Some applications of ANNs have been conducted over years in geography-related domains, such as land use and land cover change (Li and Yeh 2002; Pijanowski et al. 2002), urban modeling (Grekousis et al. 2013), remote sensing (Li et al. 2014), and spatial analysis and modeling (Govindaraju and Rao 2013; Nourani et al. 2013; Nevtipilova et al. 2014). For a detailed summary of ANN applications in geospatial science, please refer to Gopal (2017).

We conducted a bibliometric analysis on academic publications with respect to spatial modeling using neural networks (Web of Science database was used; keywords: spatial modeling and neural network). Figure 4.2a shows the results of the bibliometric analysis in terms of research domains. As we could see, the major research areas of publications are engineering, computer science, and neurosciences

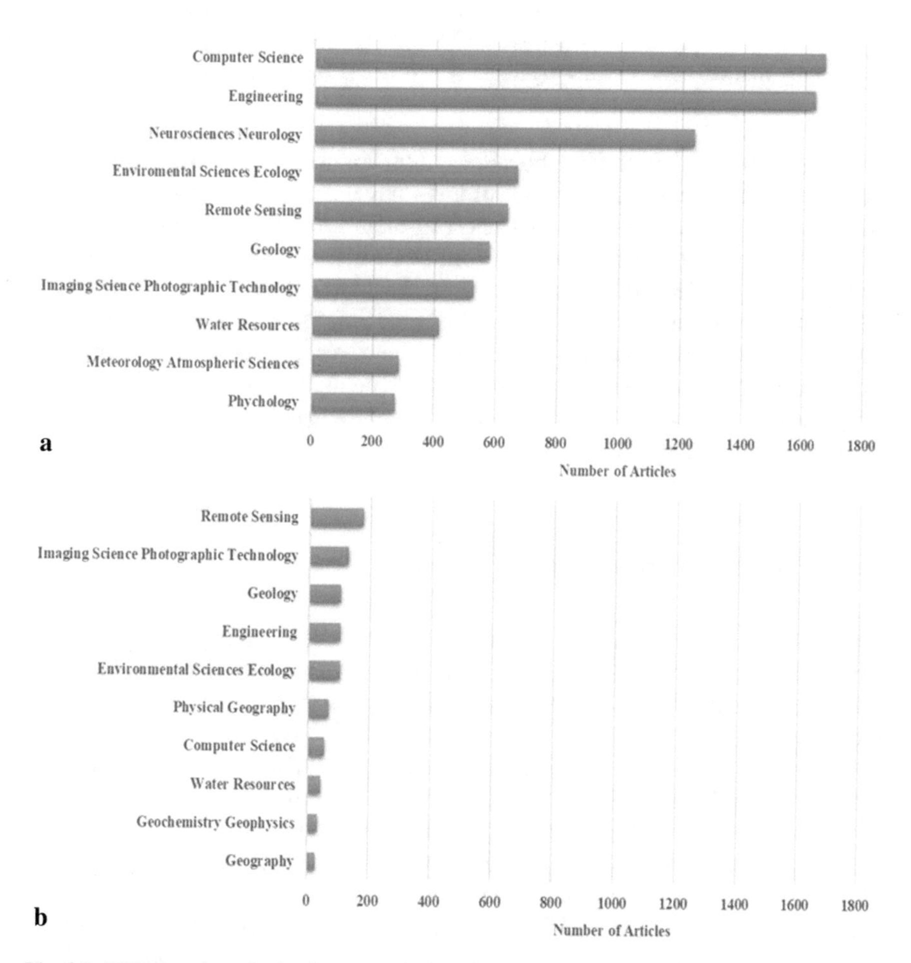

Fig. 4.2 Bibliometric analysis of research areas in terms of spatial modeling using neural networks (Web of Science database was used; all collected publications were published prior to 2018; **a**: keywords: spatial modeling and neural network, total number of articles: 6,354; **b**: keywords related to land change studies were used to refine the result of **a**, keywords: land change, land cover, land use, urban growth, intra-urban growth, or urban sprawl, total number of articles: 384; please note that a single publication may be associated with two or more domains)

Table 4.1 Summary of setting parameter method of neural network

Method	Citation
Trial and error	Heermann and Khazenie (1992), Hsu et al. (1995), Gopal and Woodcock (1996), Ozesmi and Ozesmi (1999), Rigol et al. (2001), Pijanowski et al. (2002), Fisher et al. (2003), Rigol-Sanchez et al. (2003), Erbek et al. (2004), Pijanowski et al. (2005), Nayak et al. (2006), Tang et al. (2009), Kia et al. (2012), Linares-Rodriguez et al. (2013), Isik et al. (2013), Pijanowski et al. (2014)
Literature	Miller et al. (1995), Bradshaw et al. (2002), Olden and Jackson (2002), Mas et al. (2004), Biswajeet and Saro (2007), Almeida et al. (2008), Pradhan et al. (2010)
Literature + trial and error	Berberoglu et al. (2000), Li and Yeh (2002), Joy and Death (2004)

neurology. About 6% of these publications (384) are related to land change studies (see Fig. 4.2b), to which the use case of this study belongs. Although a number of previous studies have been reported, most of them use a trial-and-error approach that depends on a set of guidelines to determine the network configuration (Stathakis 2009; Karsoliya 2012; Table 4.1). However, the trial-and-error approach is based on a manual mechanism and is often subjective. Yet, the model configurations of ANNs are dependent on the data and research objectives. The recommendation from the literature may be biased for determining the configurations of ANNs. Thus, it is necessary to resolve these limitations, which motivates us to introduce the use of hyperparameter optimization in this study.

4.2.2 Hyperparameter Optimization

Standard parameters explain the performance of a model based on a specific dataset and are determined based on the data. Hyperparameters, however, cannot be determined (or learned) from the data. Instead, hyperparameters represent higher-level properties of a model, such as the architecture of the model or model type (e.g. linear regression, ANN). As Bergstra and Bengio (2012) elicited '... a learning algorithm produces f through the optimization of a training criterion with respect to a set of parameters θ. However, the learning algorithm itself often has bells and whistles called hyper-parameters λ, and the actual learning algorithm is the one obtained after choosing λ …' (Page 1). In other words, hyperparameters are usually fixed before the training process begins and the hyperparameters could minimize generalization error of the learning algorithm. Many machine learning algorithms are associated with a set of hyperparameters. For example, the height of a decision tree, learning rates or network configuration for ANNs are representative of hyperparameters.

Hyperparameter optimization is a way of choosing a set of hyperparameters for a learning algorithm. The goal of this process is to optimize the performance of the learning algorithm on a specific dataset. Extensions and applications of the hyperparameter optimization have been conducted over the past few years by the practitioners in the field of computer science and engineering (Fig 4.3a). However, studies on optimizing hyperparameters of ANNs remain inadequate (see Fig 4.3b). Of note is the work by Thornton et al. (2013), in which they built a toolkit—Auto-WEKA to solve the combined algorithm selection and hyperparameter optimization problem (CASH) based on Bayesian optimization. But, Auto-WEKA only focuses on handling classification problem. Auto-WEKA 2.0 adds regression algorithms and supports parallel runs on a single machine to improve the computing performance (Kotthoff et al. 2016). Meanwhile, there are a set of python libraries that support hyperparameter

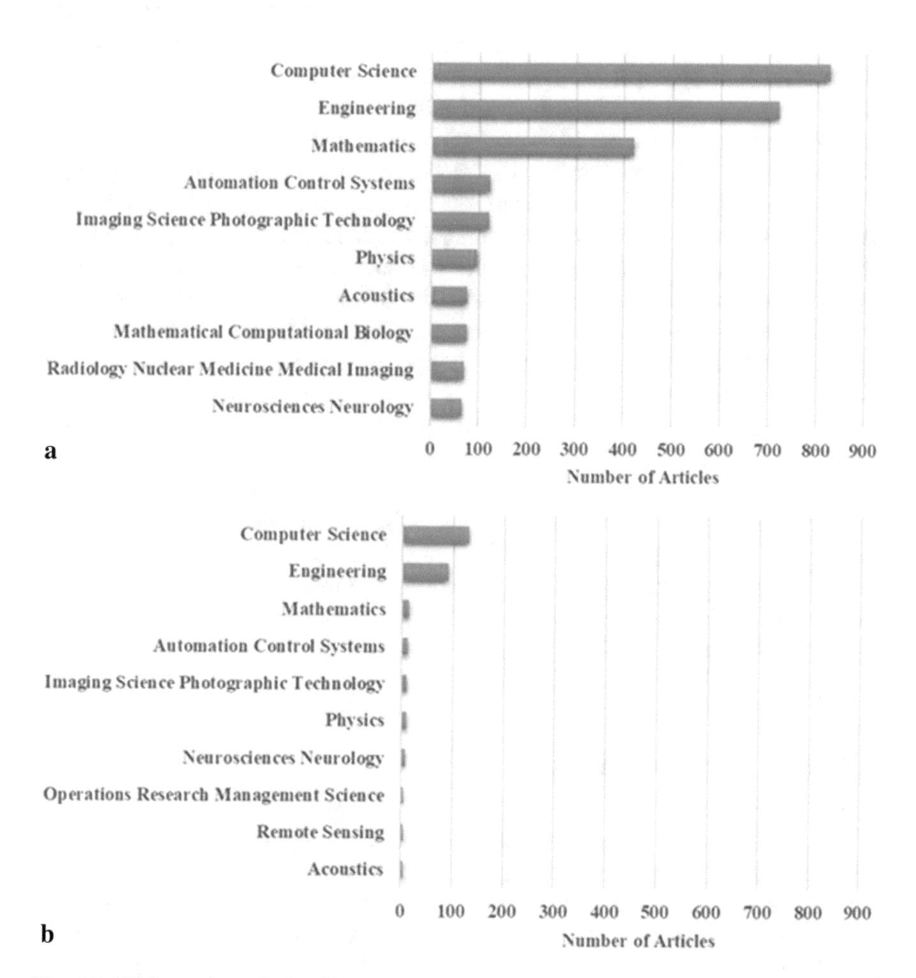

Fig. 4.3 Bibliometric analysis of hyperparameter and neural networks in terms of research areas (Web of Science database was used; **a**: keywords: hyperparameter, total number of articles: 2,115; **b**: keywords: hyperparameter and neural network, total number of articles: 275; please note that a single publication may be associated with two or more domains)

optimization with multiple machine learning algorithms, such as Hyperopt (Bergstra et al. 2015) and NeuPy (http://neupy.com/pages/home.html) (Fig. 4.3).

The general problem-solving technique for hyperparameter optimization is based on brute force search or exhaustive search, which is to find the best hyperparameter set(s) through traversing all possible combinations of hyperparameter in a given search space. Because of the higher cost and time-consuming nature of brute force search, we often use a sampling selection approach for the selection of optimal hyperparameters. Two primary selection methods support applications of hyperparameter optimization: grid search and random search. Grid search, similar to systematic sampling, manually defines a subset of the search space and then traverses all combinations of the specified hyperparameters (Lerman 1980; LaValle et al. 2004). Random search is more efficient than grid search because it randomly selects a chosen number of hyperparameters pairs from a given domain and then tests these combinations (Andradóttir 2006; Bergstra et al. 2011).

However, there exist a set of limitations of hyperparameter optimization. First, we need to evaluate the performance of a model with every combination of hyperparameters. For a single evaluation process, the computing time may be acceptable, but evaluation time is exacerbated when training multiple models. Although some applications have parallel computing capabilities or automation support for performing hyperparameter optimization (e.g. Hyperopt, Auto-WEKA 2.0), all of them focus on addressing the computational efficiency at computing level (using parallel computing here). However, it is possible that computing performance may be improved by the information from the model level, such as reducing sample size (remove redundant sampled observations here). Second, a stochastic component usually exists in machine learning algorithms. This stochasticity can be addressed via repeated measures design (Batista and Monard 2003; Kotsiantis et al. 2007), but such solution will dramatically increase the computing needs. Third, the search space of hyperparameter is usually complicated, which will further induce more computing cost. In general, the primary challenge of hyperparameter optimization is to find the optimal setting of a machine learning algorithm efficiently (e.g. the minimal amount trials) (Lerman 1980; LaValle et al. 2004).

4.3 Study Area and Data

Our study region is Mecklenburg County, North Carolina, which is located at the southwestern part of the state (see Fig. 4.4). The Mecklenburg County includes the city of Charlotte, and serves as the major area of the Greater Charlotte Metropolitan Region. This Metropolitan Region ranks the 22nd largest metropolitan area in the United States with a total population of 2,380,314 in 2014 (data source from US Census Bureau, https://www.census.gov). Charlotte is the largest city of North Carolina, and the second largest city in the southeastern U.S. Also, Charlotte is

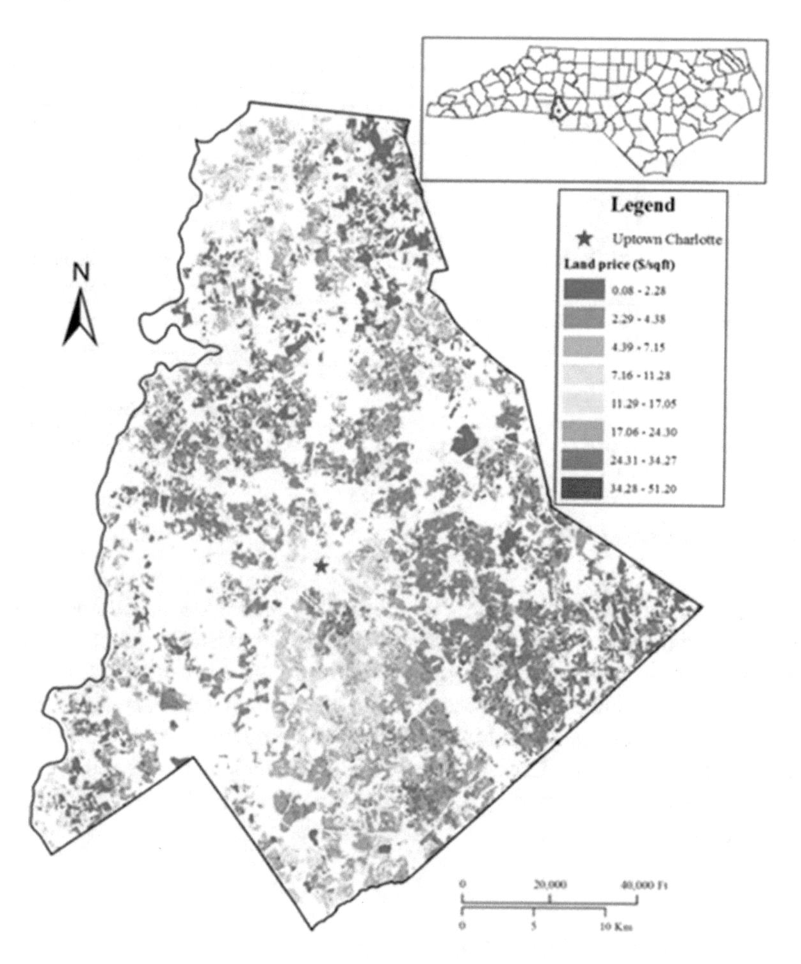

Fig. 4.4 Spatial distribution of land price of single-family houses in the study area (Charlotte, North Carolina, U.S.A)

home to the corporate headquarters of Bank of America and the east coast operations of Wells Fargo, which together with other financial institutions makes it the second-largest banking center in the United States.

In this study, we focus on investigating the impact of social environmental variables on land price. We collected our datasets from Open Mapping web portal of Mecklenburg County (see http://maps.co.mecklenburg.nc.us/openmapping). Table 4.2 reports the detail of the datasets. We used Euclidean distance when we consider the influence of spatial proximity. In this study, we explore the relationship between the land price of single-family houses and six driving factors (see Table 4.2). Figure 4.3 shows the spatial distribution of land price of single-family houses.

Table 4.2 List of datasets collected and used in this study

Dataset	Year	Name used in this study	Explanation
Land price	2016	lp	Property related information
Uptown Charlotte	2016	dis2C	Distance to city center
Park	2016	dis2P	Dataset contains Park and Recreation park
Public school	2015–2016	public_sch	Dataset contains elementary, middle, and high schools
Private school	2015–2016	priv_sch	Dataset contains elementary, middle, and high schools
Road	2016	dis2R	Dataset contains developed interstate highways
Hospital	2016	hospital	Locations of Hospitals

4.4 Methodology

Land price evaluation and hyperparameter optimization are the main components of the automated model selection for ANN-driven spatial model. One of our objectives in this study is to investigate how hyperparameters influence the performance of the ANN model. Another objective is to improve the computational efficiency using other approaches together with high performance computing. Also, we will give a recommended hyperparameter set(s) for the optimal model. In this section, we discuss the land price evaluation model and the algorithm of ANN-driven hyperparameter optimization and then illustrate the implementation.

4.4.1 Land Price Evaluation Model

We developed a land price evaluation model to investigate the relationship between the land price in the study region and their drivers. We focus our study on the land price of residential housing. Residential housing, as one section of the real estate market, is one of the principal components of our daily life. Real estate market is an essential part of the regional economy. The economic crisis in 2008 has shown that the real estate market is affected by such factors as employment, financial system stability, policies and so on. The relationship between the real estate market and the economy has been studied in the literature (Quan and Titman 1999; Girouard and Blöndal 2001; Quigley 2002; Mera and Renaud 2016). Some studies suggested that the determinant of residential land price may include accessibility value, amenity value, and topography (Brigham 1965; Arribas et al. 2016).

Thus, in this study we chose six variables as drivers of land price in our study region: (1) distance to the center city of Charlotte (noted as x_1), (2) distance to the nearest interstates (x_2), (3) distance to the nearest public school (includes elementary school, middle school, and high school) (x_3), (4) distance to the nearest private school (includes elementary school, middle school, and high school) (x_4), (5) distance to the nearest park (x_5), and (6) distance to the nearest hospital (x_6). That is, our land price evaluation model can be formulated as:

$$p = f(x_1, x_2, \ldots, x_n) \tag{4.1}$$

where p is the land price of a parcel. x_1–x_n are influential variables as drivers of the land prices associated with the parcel ($n = 6$ here). *f(.)* represents the model that establishes the relationship of land price with its drivers. Thus, for linear regression approach, the land price evaluation model in this study becomes the following:

$$p = w_0 + w_1x_1 + w_2x_2 + w_3x_3 + w_4x_4 + w_5x_5 + w_6x_6 \tag{4.2}$$

where w_0–w_6 are weights (or coefficients) that define the linear regression model for land price evaluation. These weights can be estimated by ordinary least squares (OLS), maximum likelihood estimation (MLE) or other estimation methods. x_1–x_6 are influential variables as drivers of the land prices associated with the parcel.

If we use the neural network approach, the land price evaluation model *f(.)* can be represented by a neural network with one hidden layer as below:

$$p = f_a^{out}\left(\sum w_{ij}^1 f_a\left(\sum w_{ij}^0 x_n\right)\right) \tag{4.3}$$

where f_a^{out} is the activation function for output layer neuron(s), f_a is the activation function for neurons in hidden layers. $w_{ij}^{l=0,1,2..}$ is the weight between node i in layer l to and node j in layer $l + 1$. w_{ij}^0 stands for the weight between the first two layers (i.e., the input layer and the hidden layer). w_{ij}^1 means the weight between the second and the third layer (e.g., the hidden layer and the output layer here).

Further, the land price evaluation model *f(.)* also can be written as a two-hidden-layer neural network as in Eq. 4.5:

$$p = f_a^{out}(\sum w_{ij}^2 f_a(\sum w_{ij}^1 f_a(\sum w_{ij}^0 x_n))) \tag{4.4}$$

where w_{ij}^1 is the weight between the first hidden layer and second hidden layer, and w_{ij}^2 means the weight between the second hidden layer and the output layer.

Weights of neural networks can be regarded as standard parameters of the neural network model. Thus, the linear regression model is a special type of neural network with linear activation function. The parameters of learning algorithms used to generate these weights become hyperparameters.

4.4.2 *Hyperparameter Optimization*

Figure 4.5 illustrates the framework of hyperparameter optimization in this study. The framework consists of four major phases: (1) generation of hyperparameters, (2) acceleration of model runs using parallel computing (discussed in detail in Sect. 4.4), (3) evaluation of sampled hyperparameters (discussed in detail in Sect. 4.3), and (4) analysis of hyperparameters-derived results (Fig. 4.5).

In the first phase of generation of hyperparameters, we select the examined hyperparameter space and then select sampled hyperparameters based on selection methods (e.g. grid search and random search). There are a number of hyperparameters that have impacts on ANNs, such as batch size, training iterations, and the number of

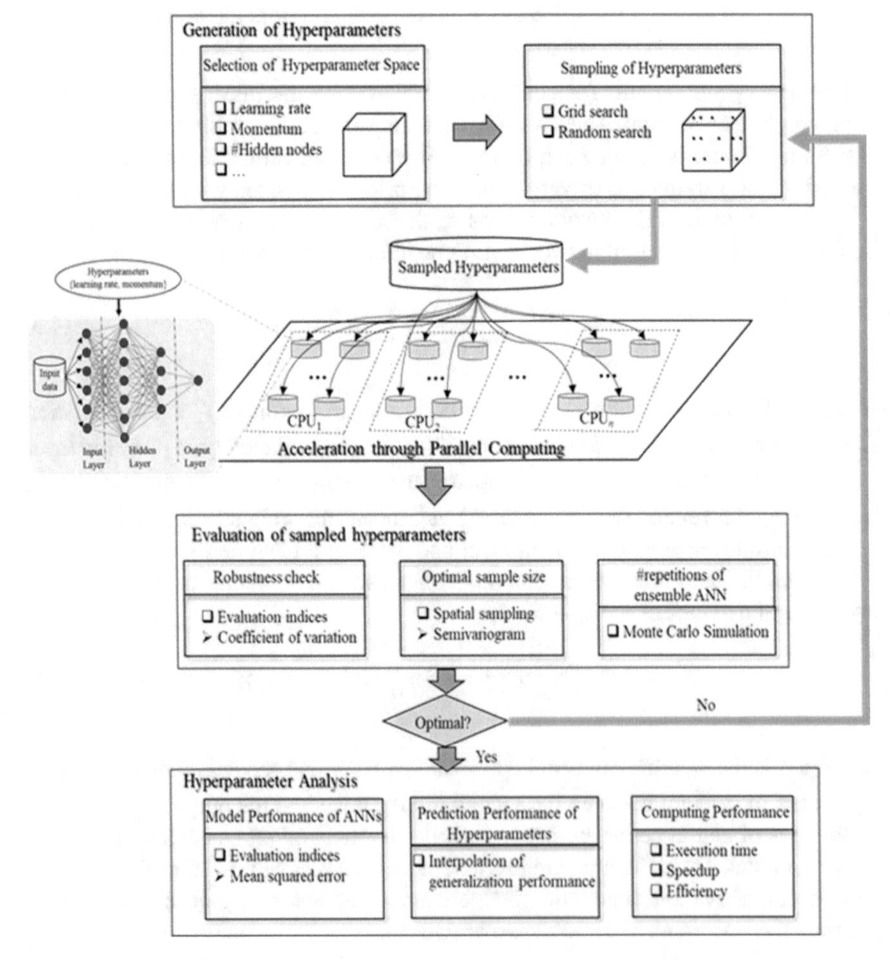

Fig. 4.5 Framework of the use of hyperparameter optimization for artificial neural networks (ANNs)

hidden units. In this study, we focus on examining two representative hyperparameters for backpropagation feed-forward neural network modeling: learning rate and momentum. The updated weight formula (Eq. 4.6) is listed as follows:

$$\Delta w_{ij}^{l}(n) = -\delta \frac{\partial E(n)}{\partial w_{ij}^{l}(n)} + \alpha \Delta w_{ij}^{l}(n-1) \tag{4.5}$$

where δ is the learning rate and α is the momentum term. Δw_{ij}^{l} is the adjusted weight between the target output for the current training example and the output generated using gradient descent rule—$\frac{\partial E(n)}{\partial w_{ij}^{l}(n)}$. n is the step of current iteration (or epoch).

In Sect. 4.2.2, we discussed the selection methods of hyperparameter optimization. In this study, we used both grid search and random search for comparison purpose. For random search, we used Latin hypercube sampling (LHS) approach (see (McKay et al. 1979). LHS first divides the entire search space into a set of Latin squares, then randomly identify one or more sample points in each square. In this study, we investigate the efficiency and accuracy of these two sampling methods using ANNs with two hidden layers. Figure 4.6 shows the distribution of the sampled hyperparameters (learning rate and momentum) for grid search and random search (more detail provided in Results Section).

Regarding the analysis of hyperparameters (Phase 4 in the framework; see Fig. 4.5), it includes three modules: (1) model performance analysis of ANNs, (2) prediction performance analysis of hyperparameters, and (3) computing performance. For the first module, mean squared error (MSE) is applied as model performance metric here. For continuous variables, MSE is a performance metric that has been extensively used (Willmott 1982; Brown 1998). Performance metrics based on the confusion matrix (e.g. overall accuracy, and Kappa coefficient; see (Fawcett 2006)) are suitable for categorical variables. In our case, given that land price is a continuous variable, the error function that we used for performance evaluation is MSE (see Eq. 4.7),

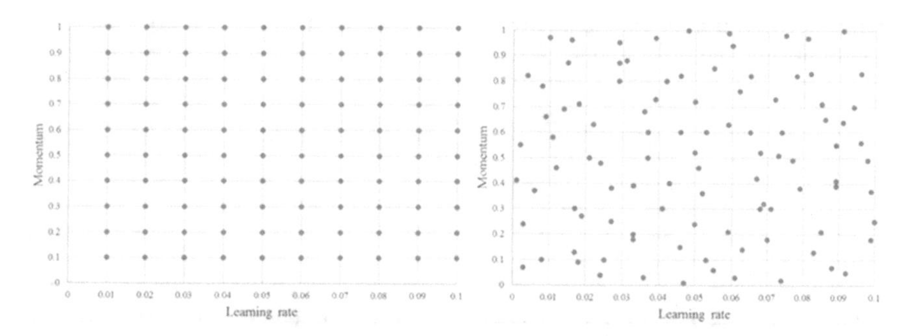

Fig. 4.6 Distribution of sampled hyperparameters (**a**: grid search; **b**: random search)

$$MSE = \frac{1}{n}\sum_{i=1}^{n}\left(\widehat{p_i} - p_i\right)^2 \quad (4.6)$$

where $\hat{p}_i$ is the estimated or predicted land price of the ith data record. p_i is the observed value of the ith data record. n is the total number of records of the dataset. In this study, we also use the coefficient of variation (CV) to investigate the generalization performance of the model (Phase 2 in the framework; see Fig. 4.5). CV is a measure of relative variability that is comparable across different parameter sets (Brown 1998). The formula for CV is:

$$CV = \frac{\delta}{\mu} \times 100\% \quad (4.7)$$

where δ is the standard deviation of a variable. μ is the average of the variable. Typically, a model or an algorithm is robust when CV is less than 100% (i.e., when standard deviation equals the average).

4.4.3 Determining Optimal Sample Size

For the evaluation of selected hyperparameters, we involved spatial sampling approach. Sampling is to acquire a certain number of samples instead of estimating characteristics of the whole population due to higher cost and longer computing time for obtaining information from the entire population. Classical sampling theory assumes that data are identically and independently distributed, which is often referred to as design-based sampling strategy (Särndal et al. 1978). In other words, in the design-based sampling strategy, the sample size is fixed and the sampling locations are random. Sample size determination formulas based on classical sampling theory were proposed (shown below in Eqs. 4.9 and 4.10). Equation 9 is to determine sample size for a finite population (Krejcie and Morgan 1970), and Eq. 4.10 is for the infinite population (Godden 2004):

$$s = \frac{\chi^2 Np(1-p)}{d^2(N-1)} + \chi^2 p(1-p) \quad (4.8)$$

$$n = \frac{z^2 p(1-p)}{d^2} \quad (4.9)$$

where s is sample size for a finite population based on the chi-square distribution, and n is for an infinite population which follows a normal distribution. χ^2 is the chi-square for 1 degree of freedom at the significant level of 0.05 (3.841 here), whereas z is the critical value at confidence level (e.g., 1.96 for 95% confidence level). N is population size and p denotes the population proportion. d as the degree of accuracy (i.e., margin of error) expressed as a proportion. They assumed p is 0.5 since this would provide the maximum sample size, and d is set to 0.05.

A given sampling point may influence other points located close by, or even some distance away (Tobler 1970). Spatial sampling theory (aka, model-based sampling strategy), which is based on geostatistics, is to address this form of intersample influence. From a geostatistical perspective, the existence of spatial dependence in the search space implies that the independence assumption for classical sampling is not met. Spatial dependence occurs when information available at one location allows us to infer information about the other location. However, besides related fields (e.g. ecology, geography, evolutionary biology), the consideration of spatial dependence has not been discussed in other fields. But, the spatial dependence can be found in georeferenced data or spatiotemporal data (Legendre and Fortin 1989). Moreover, the field of geostatistics provides a set of approaches (e.g., semivariogram) to estimate the spatial dependence and use this information to predict unknown locations under a given study.

The semivariogram model describes the spatial dependence of the measured sample points. There are three characteristics of semivariogram: range, sill and nugget (Matheron 1963). To provide an efficient way to consider sample size with spatial dependence, Griffith (2005) proposed a formula for estimating effective sample size[1] (Module 2 of Phase 3 in the framework; see Fig. 4.5) using semivariogram models:

$$n^* = \frac{ni}{ni - \frac{C_1}{C_0+C_1}\sum_{i=1}^{ni}\sum_{j=1, j\neq i}^{ni} f\left(d_{ij}, r\right)}n \tag{4.10}$$

where n^* is the independent sample size, and ni is the spatially autocorrelated sample size (refer to the sample size from Eq. 4.9 or Eq. 4.10). $f(d_{ij}, r)$ stands for a particular semivariogram model with range (r), nugget (C_0), and slope (C_1, C_1 = sill/range), and d_{ij} is the distance between location i and j. The following equations are two model-specific cases when C_0 is zero:

$$\text{spherical}: \frac{ni - 1}{\left(1 + 251.5132\frac{r}{d_{\max}}\right)^{1.9324}} + 1, d_{ij} \leq r \tag{4.11}$$

$$\text{exponential}: \frac{ni - 1}{\left(1 + 51.4879\frac{r}{d_{\max}}\right)^{1.757}} \tag{4.12}$$

where $d_{\max}$ denotes the maximum distance between location i and j. Comparing with the equation for statistical approach, semivariogram approach considers (1) the types of semivariogram model (e.g., spherical, exponential, and gaussian), and (2) the

[1] Effective sample size (aka, adequate sample size) is a notion defined for a sample that is statistically significant and the observations in the sample are correlated.

spatially autocorrelated sample size, which is estimated from a statistical approach (Eqs. 4.9–4.10). Thus, we can use effective sample size to adjust the experimental design and improve the computing performance.

However, a hyperparameter search space may contain an infinite number of combinations of hyperparameter sets, and these sampled hyperparameters can only represent part of the hyperparameter space. Thus, we further use prediction methods to estimate the generalization performance of the entire hyperparameter space. There exist a set of interpolation methods (Krige 1978; Zimmerman et al. 1999), such as kriging and inverse distance weighting (IDW). The difference between kriging and IDW is that the former allows for the consideration of spatial dependence but the latter does not. In this study, we used kriging to predict and delineate the continuous patterns of CVs and MSEs from those of sampled hyperparameters. Kriging is a spatial interpolation method that makes use of semivariogram to calculate the spatial dependence between points at different lag distances. To address computational efficiency (Module 3 of evaluation of sampled hyperparameters in the framework; see Fig. 4.5), Monte Carlo approach (Kalos and Whitlock 2008) was applied in this study. Monte Carlo approach can be used as significance testing approach. That is, the number of Monte Carlo approach required is statistically significant or not, thus allowing for determining the minimum number of repetitions of neural network runs. The minimum number of repetitions (n_i) at a particular confidence interval is shown in Eq. 4.14:

$$n_i = \left[\frac{zS_x}{E\overline{x}}\right]^2 \tag{4.13}$$

where z is the critical value of confidence level (e.g., 1.96 for 95% confidence level). S_x denotes the sampled standard deviation, and E is the margin of error (i.e., 0.05 in the case). Sample mean assigns to $\overline{x}$.

In addition, data normalization is used to avoid unnecessary computation here. The entire dataset is split into a training dataset and a testing dataset for supervised learning. In our model, we used linear transformation approach to standardize the input data to the range of [0, 1] (see Eq. 4.15).

$$x_{new} = \frac{x - x_{\min}}{x_{\max} - x_{\min}} \tag{4.14}$$

where x_{new} is the normalized value of a specific record for a variable. x is the original value of the record. $x_{\min}$ and $x_{\max}$ are the minimum and maximum of the variable. Output data will then be scaled back.

4.4.4 Parallel Computing and Implementation

In Phase 2 of the framework (see Fig. 4.5), we illustrate the acceleration through parallel computing because the computational demand for hyperparameter optimization is heavy. In this study, we have n hyperparameter sets each of which needs m times of repetitive runs. We request k CPUs from high performance computing cluster and organize the set of all runs (m × n) into k subsets. Further, we could deploy k sub-sets of model runs into k CPUs—that is, each CPU runs mn k jobs. For example, if there are n = 100 hyperparameters sets each repeating m = 100 runs over k = 100 CPUs, then the number of jobs is 100 and each CPU charges a single job. Because there is no communication among model runs, the parallel computing approach is the so-called embarrassingly parallel computing (Wilkinson and Allen 1999; Tang and Jia 2014). To evaluate parallel computing performance (Module 3 of Phase 3 in the framework; see Fig. 4.5), we use speedup (*sp* in Eq. 4.16) and efficiency (*e* in Eq. 4.17) for evaluating the acceleration of parallel computing.

$$sp = \frac{t_1}{t_n} \tag{4.15}$$

$$e = \frac{sp}{n_{cpu}} \tag{4.16}$$

where t_1 is the sequential computing time using a single CPU and t_n is the parallel computing time using n CPUs. n_{cpu} is the number of CPUs used for acceleration. Speedup uses to calculate the acceleration rate. Theoretically, the ideal speedup (aka, linear speedup) is n with n CPUs. Efficiency is the ratio of speedup over the number of CPUs used for parallel computation (Wilkinson and Allen 1999).

The high-performance computing cluster that we used has 59 nodes connected through a gigabit network to accelerate the automated hyperparameter optimization process. Each computing node of the high-performance computing cluster has 12 CPUs and 36 GBs of memory, in total 432 CPUs (dual Intel Xeon 2.93 GHz). The operating system of the cluster is Redhat Linux, and the job scheduling system is Torque/PBS. Shell scripts were used to wrap the computation of hyperparameter optimization into computing jobs that can be executed on the Linux cluster in parallel. The neural network library that we used in this study is FANN (http://leenissen.dk/fann/wp/), which was written in C programming language. Further, Python scripting language was used for data processing associated with neural network modeling and GIS analysis.

4.5 Results

In this study, we focus on examining the utility of hyperparameter optimization for spatially explicit modeling driven by ANNs. We organized the entire dataset into 32 groups based on the distance to the center city of Charlotte (see Fig. 4.7; the distance interval is 1 km). There are two reasons that we used distance to the center city of Charlotte. First, cities typically expand outward from the center over time in urban economics, and the center of a city is usually called downtown or uptown. When a parcel becomes suitable for development as the city expands, the parcels with similar distances may have similar start dates regarding development (Cunningham 2006). Second, a number of studies reported that the distance to downtown or uptown has a substantial influence on land price (Heikkila et al. 1989; Atack and Margo 1998; Yamazaki 2001; Hu et al. 2016) (Fig. 4.7).

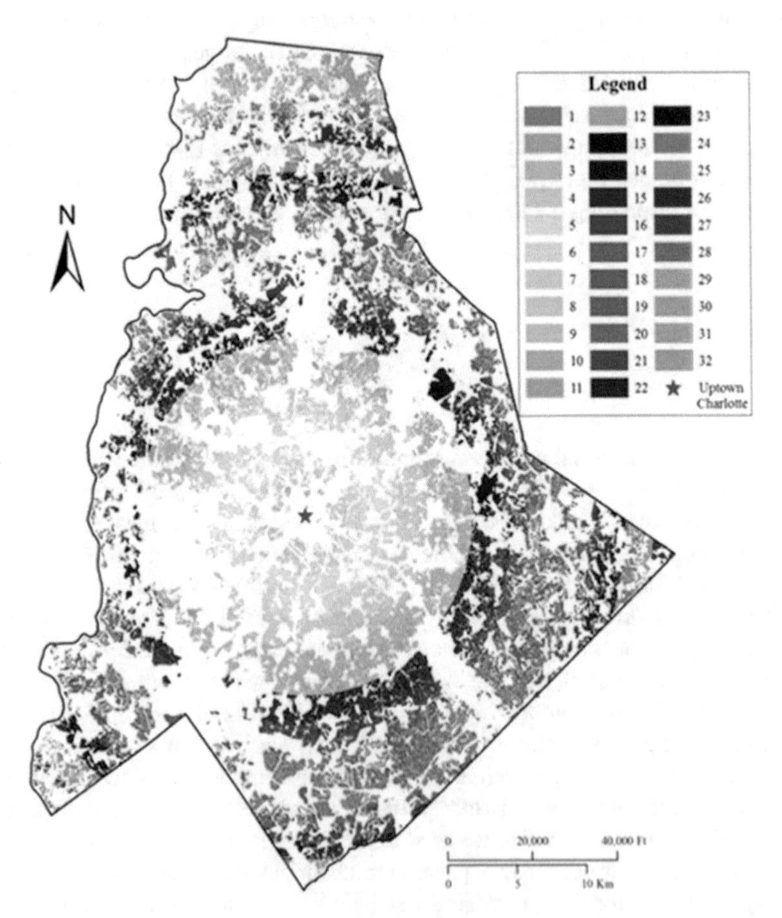

Fig. 4.7 Spatial pattern of single-family housing parcels in terms of distance to uptown Charlotte in the study area (Mecklenburg County, NC; city: Charlotte)

Table 4.3 Results of linear regression modeling (see Table 4.2 for definitions of these variables)

Variable	Coefficient	Standard error	t-statistic	Probability
Intercept	16.95	0.9359918	18.11	0.00
dis2P	0.004	0.0001177	35.95	0.00
dis2C	0.0006	0.0000743	7.97	0.00
dis2R	0.00008	0.0000462	1.78	0.074
Public_sch	−0.001	0.0000640	−14.91	0.00
Priv_sch	−0.00005	0.0000338	−1.61	0.108
Hospital	−0.0015	0.00001296	−113.43	0.00
R^2	0.7183			
Adjusted R^2	0.7180			

Specifically, we chose one of the groups as our experimental dataset. In this case study, group 4 with 9254 records (parcels here) was selected: 6478 records (70%) were assigned as the training set, and the remaining 2776 records (30%) for testing.

We used the same training dataset to fit the linear regression model and the same testing dataset was used to predict values based on the fitted linear model. The ordinary least squares model was applied in this study. Table 4.3 summarizes the results of the linear regression model. The R^2 of the linear model is 0.7183, and the predicted MSE based on the testing dataset is 0.0068. In the ANN model, the MSE of the optimal hyperparameter set is around 0.004 for both hyperparameter methods (reported in detail next). These results suggest that the ANN model has a better generalization performance than linear regression modeling.

In this study, ANNs were configured as two hidden layers with six nodes in the first layer and three nodes in the second layer. We applied sigmoid-type activation function (Eq. 4.18) for hidden layer neurons and linear identity activation function for output layer (Eq. 4.19).

$$f_a(in) = 1/\left(1 + e^{-in}\right) \tag{4.17}$$

$$f_a^{out}(in) = in \tag{4.18}$$

where *in* is the weighted sum of values of neurons from the previous layer.

Model fitting is an important part of all quantitative models. However, because of the complex nature of ANNs, overfitting is a serious and common problem in training ANNs. There exist a set of strategies for handling the overfitting issue, such as cross-validation, early stopping, or ensemble learning techniques. In this study, we used testing dataset to detect whether the model has overfitting issue or not. We found that most learning processes have overfitting issue when the number of iterations reaches to 10,000 or after. Also, the minimal MSEs are between 0.003 and 0.004. Thus, the stopping criteria are based on the number of iterations (10,000 epochs) and the threshold of MSE (0.003 here) to avoid overfitting issue. In other words, when the training MSE is close to the threshold (0.003 here), or the learning algorithm reaches the number of iterations (10,000), the learning process will stop. We repeat each hyperparameter set 100 times, and the initial weights of each ANN set to 0.00.

4.5.1 Results of Grid Search and Random Search

The interval of learning rate and momentum of our search space set to 0.01 and 0.1. Range of learning rate and momentum are (0, 0.1] and (0, 1]. We applied 100 sampled hyperparameter sets in this study. For grid and Latin hypercube sampling, we created a 10 × 10 square grid—i.e., the total number of square grids is 100.

4.5.1.1 Model Performance

We used the testing MSEs to evaluate the ANN-based model performance for sampled hyperparameters. Figure 4.8 illustrates the scatterplot of averaged testing

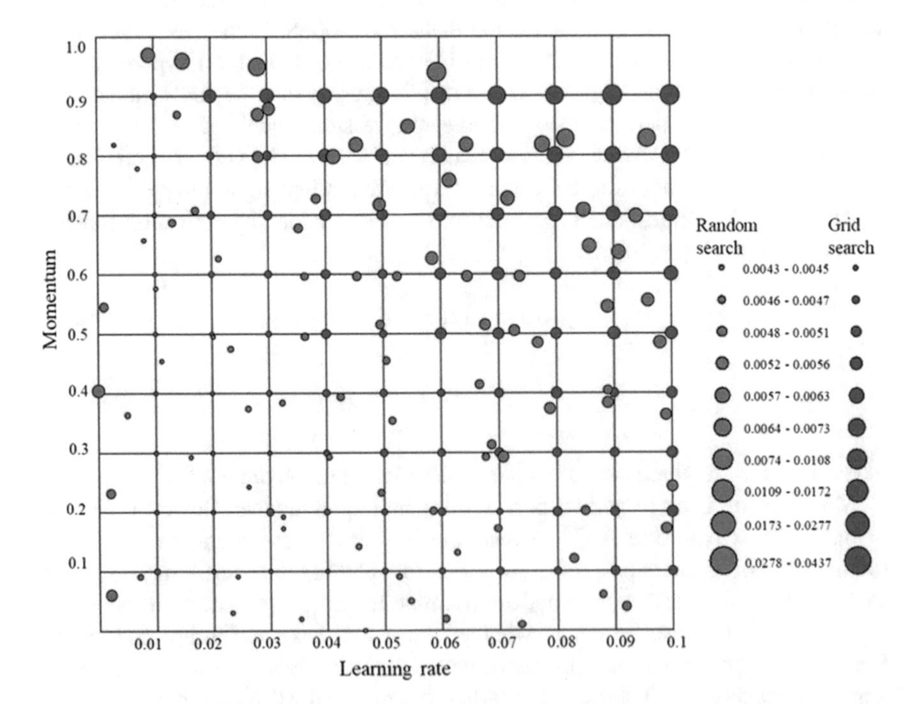

Fig. 4.8 Scatterplot of averaged testing MSEs for hyperparameter sets (circle size is proportional to MSE, circle size from small to large stands for low MSE to large MSE; outliers were excluded)

MSEs (generalization performance) for grid search and random search (outliers were excluded). The effective hyperparameter sets for grid search is 90, whereas 94 for random search. For grid search, high testing errors concentrate around the regions where momentums are large. In particular, testing errors reach the maximum when momentum is around 0.9. While testing errors are also affected by learning rates: testing errors tend to increment as learning rates increase. Similar patterns can be observed from the scatterplot of random search while the parameters sets are randomly generated. In general, generalization performance of neural networks is high (i.e., low testing MSEs) when both learning rates and momentum are low. Meanwhile, low learning rates coupled with relatively large momentum (e.g., momentum <0.7) have high generalization performance.

Figure 4.9 shows the scatterplot of CVs for both grid and random search with outliers removed. We calculated CV based on the result of the testing dataset. High CVs occur at those regions where learning rate and momentum are large. The highest CVs (>100%) appear when momentum is larger than 0.9 for random search, whereas learning rate is higher than 0.07 for grid search. However, there exists an outlier (CV > 100% here) in grid search when learning rate is 0.03, and momentum is 0.2. But, the averaged testing MSE is small in this hyperparameter set (see Fig. 4.6). Similarly, CVs from random search show relatively small values around this region.

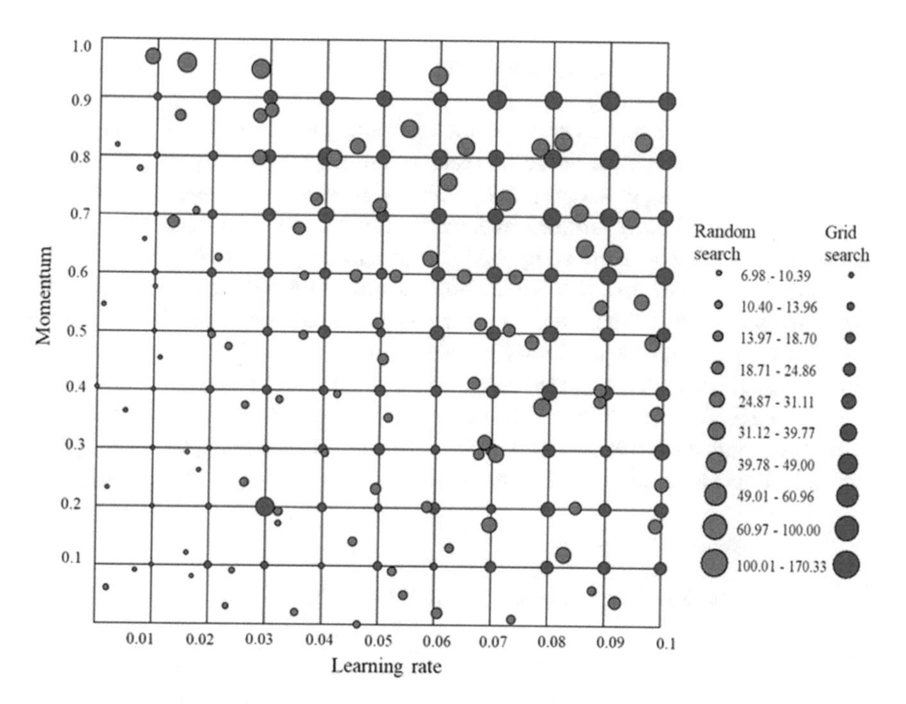

Fig. 4.9 Scatterplot of CVs for hyperparameter sets (circle size is proportional to CV, circle size from small to large stands for low CV to large CV; outliers were excluded)

4.5.1.2 Optimal Sample Size

As we discussed in Sect. 4.3.3, semivariogram is a spatial statistical approach that can be used to estimate spatial dependence. However, in most cases, how to determine the semivariogram would be a major part of spatial sampling. If the semivariogram is not known, we should select some samples on a regular grid (e.g., triangular, hexagonal) to create the semivariogram model (McBratney and Webster 1983). Hence, we created the spherical semivariogram model based on 100 sampled hyperparameter sets (see Fig. 4.6 for distribution of sampled hyperparameter sets).

In semivariogram analysis, the number of lags is set to 12 and lag size is 0.048. From Fig. 4.10, the nugget for both random search and grid search is 0, and the range is about 0.153 for random search and around 0.148 for grid search. This means for any two sampled hyperparameters, there is no spatial autocorrelation or dependence when the distance between the two sets is longer than 0.153 (random search) or 0.148 (grid search). The sill of random search (1.07×10^{-7}) is smaller than grid search (9.676×10^{-6}). In other words, the variance of sampled hyperparameter sets in random search is lower than that from grid search. Furthermore, according to Eq. 4.12, we only need 88 sampled hyperparameter sets instead of 383 from conventional statistics (based on an assumption of independence among samples; see Eq. 8.10).

4.5.1.3 Optimal Number of Repetitions

In this study, we focus on investigating the optimum number of repetitions for each sampled hyperparameter set based on results from the treatment group of random search (Fig. 4.10). Figure 4.11 reports estimation results of a number of repetitions (using Eq. 4.14) for each sampled hyperparameter set over a continuous hyperparameter space. Observed from Fig. 4.11, when learning rate is less than 0.05 and momentum is smaller than 0.7, the number of repetition for each sampled hyperparameter sets is lower than 100. As learning rate becomes large, the number of repetitions tends to increase. Similarly, larger momentum leads to more repetitions. However, momentum has more influence on the number of repetitions than learning rate. The maximum number of repetitions in this hyperparameter space is over 1,200 when momentum is larger than 0.9, whereas the minimal number of repetitions is less than 20 when learning rate is less than 0.01 and momentum is lower than 0.4.

4.5.2 Prediction Performance of Hyperparameters

Given the testing MSEs of sampled hyperparameters, we further used spatial interpolation (based on universal kriging) to produce a continuous surface of MSEs and CVs for the prediction of the generalization performance of hyperparameters. When spatial dependence exists in the model, a kriging-based spatial interpolation algorithm is suitable for predicting the general performance. Figure 4.12 shows the predic-

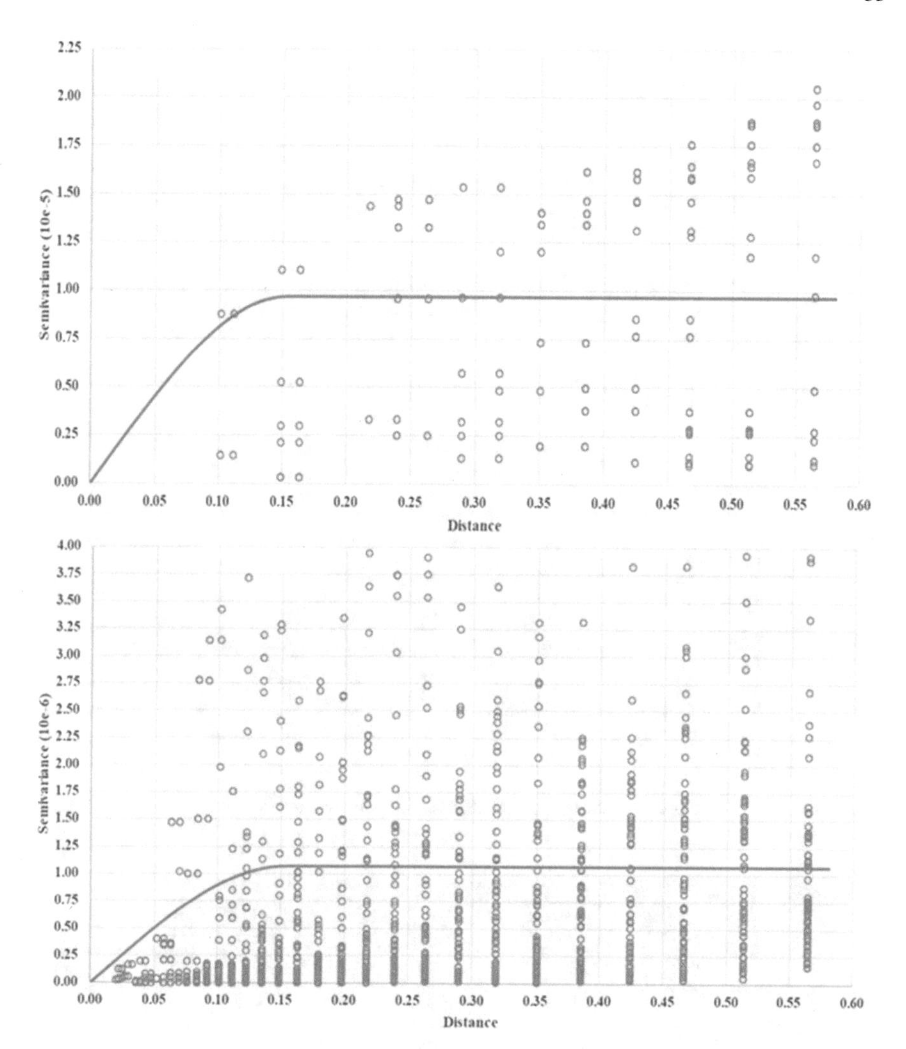

Fig. 4.10 Semivariogram analysis of grid and random sampling approach (A, grid search; B, random search)

tion maps of testing MSEs for both grid search and random search. In this prediction map of grid search (Fig. 12a; MSE of cross-validation: 1.936×10^{-5}), larger predicted errors occur as learning rate increases. Likewise, large momentum (above 0.7) also leads to high prediction error. Small learning rate and small momentum can obtain the lower predicted error. The largest error appears when momentum factor is within the range of 0.8–0.9 and learning rate is within 0.08–0.1.

Figure 12b shows the prediction error map for the random search approach. As we could observe, prediction error will increase (MSE of cross-validation: 6.76×10^{-6}) with increments in learning rates. Likewise, large momentum causes high

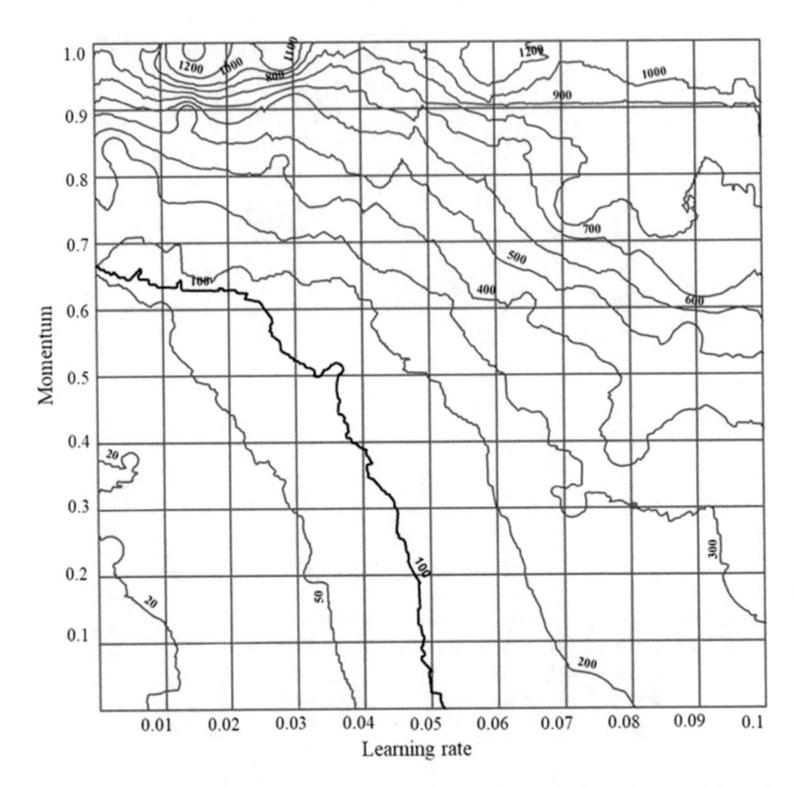

Fig. 4.11 Map of the optimal number of Monte Carlo repetitions for neural network-based spatial modeling

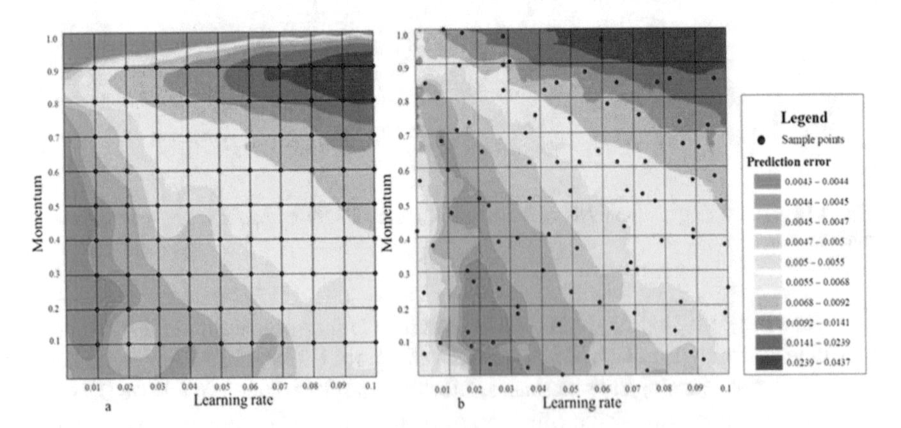

Fig. 4.12 Maps of prediction performance using hyperparameters optimization (a: grid search; b: random search; prediction error: mean squared error)

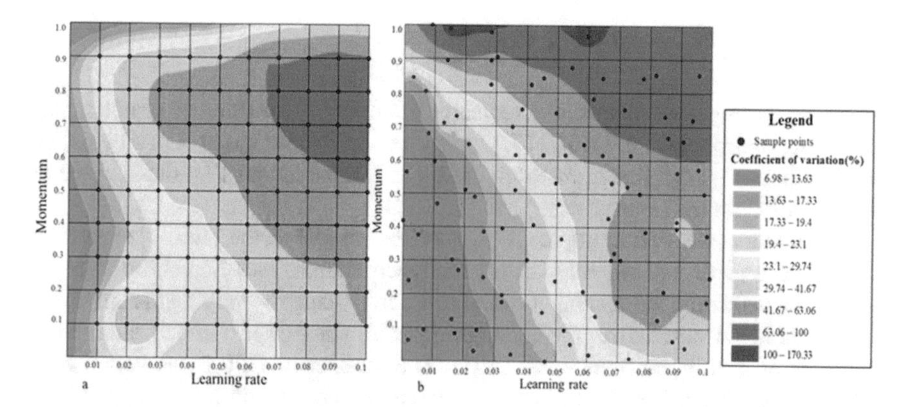

Fig. 4.13 Distribution of coefficient of variation based on testing MSEs using hyperparameter optimization methods (**a**: grid search; **b**: random search)

prediction error when learning rate is above 0.03. The general patterns of prediction error maps for both grid search and random search are similar: small learning rates and small momentum lead to high generalization performance. This generalization performance tends to be lower while both learning rates and momentums become larger. That is, the generalization performance for both grid search and random search is consistent in general. In addition, these two methods obtain different results when learning rate is less than 0.02 (left-most in Fig. 12a, b). Grid search obtains small prediction error in this region, but random search shows a relatively large error.

Figure 4.13 illustrates the maps of continuous patterns of CVs in response to learning rates and momentums. The value of CV was calculated based on testing MSEs for each sampled hyperparameter sets. For grid search, the associated MSE for cross-validation is 374.04, whereas 343.36 for random search. As we could observe, the results of CVs indicate that the hyperparameter optimization for both grid search and random search are robust (CV $\leq$ 100% for most of the hyperparameter space). The generalization performance is becoming worst when learning rate and momentum are large. From Fig. 13a, largest CVs occur when learning rate between 0.07 and 0.1 and momentum between 0.5 and 0.9. When learning rate is 0.01, the hyperparameter optimization is the most stable. Random search has a similar pattern: small learning rate and small momentum lead to stable generalization performance (Fig. 13b). However, the lowest CVs concentrate in smaller learning rate (i.e., learning rate <0.03). The generalization performance of the hyperparameter optimization approach for both grid search and random search tends to be more stable when learning rates and momentums are smaller (Fig. 4.13).

4.5.3 Parallel Computing Performance

In the computing performance, the sequential computing time for grid search hyperparameter optimization is 1,464,177.77 s (about 16.95 days; based on a single CPU on the Linux cluster). Random search used 1,157,428 s (about 13.4 days; based on a single CPU on the Linux cluster).

In this study, we used multiple CPUs to run our model on the Linux cluster. Each job is in charge of 100 ensemble runs of a hyperparameter set. Scalability analysis allows for evaluating the change of computing performance by varying the number of CPUs. Table 4.4 summarizes the computing performance of grid search and random search over a different number of CPUs. The number of CPUs was varied from 10 to 100 at an interval of 10. For grid search, the computing time dropped from 1,464,178 s (about 407 h) to 16,490 s (about 4.06 h) when 100 CPUs were used. As we could observe, computing time tends to decrease substantially when more CPUs were introduced. Similarly, speedup shows an increasing pattern with increment in the number of CPUs (from 9.91 to 88.79). Random search exhibits similar patterns as grid search. The sequential computing time for random search was 1,472,122 s

Table 4.4 Summary of computing performance of hyperparameter optimization over different number of CPUs (time unit: seconds)

	#CPUs	Parallel time	Speedup	Efficiency	Sequential time
Grid search	10	135,803	10.78	1.08	1,464,178
	20	74,860	19.56	0.98	
	30	46,157	31.72	1.06	
	40	44,549	32.87	0.82	
	50	31,063	47.14	0.94	
	60	32,100	45.61	0.76	
	70	31,525	46.44	0.66	
	80	29,146	50.24	0.63	
	90	16,490	88.79	0.99	
	100	16,490	88.79	0.89	
Random search	10	135,504	10.86	1.09	1,472,122
	20	79,341	18.55	0.93	
	30	53,376	27.58	0.92	
	40	41,614	35.38	0.88	
	50	38,839	37.90	0.76	
	60	40,518	36.33	0.61	
	70	34,002	43.30	0.62	
	80	27,785	52.98	0.66	
	90	24,950	59.00	0.66	
	100	24,950	59.00	0.59	

(about 409 h). When 100 CPUs were involved, the computing time decreased to 24,950 s (about 6.93 h). Correspondingly, the speedup increased from 8.56 to 59 with the recruitment of more CPUs.

4.6 Discussions

4.6.1 Necessity of the Framework

The configuration of neural networks has a major influence on the model performance. Hyperparameter optimization provides a potential solution to resolving the limitations associated with the nature of neural networks (i.e., complex configuration). However, the existing literature or applications mainly focused on improving the computational efficiency at the computing level, and ignored the (spatial) dependence in the hyperparameter space and potential benefits that may further improve the computing efficiency. Therefore, it is necessary to develop a framework that allows for further enhancing the computational efficiency by taking into account information at the model level. At the same time, this framework provides support for revealing the dependence in the hyperparameter space, which can be leveraged for further improving the hyperparameter optimization.

4.6.2 Feasibility of the Framework

In the scatterplots of average testing MSEs and CVs for sampled hyperparameter sets (see Figs. 4.8, 4.9), smaller learning rates and momentums resulted in better generalization performance for both grid search and random search in our case study. A large learning rate can accelerate the training process when the search crosses a plateau. However, the training of neural networks may not converge or even diverge when the learning rate is too high because the search jumps over steep regions or moves into undesirable regions (Yu and Chen 1997). Momentum can avoid local minima issue and accelerate the training process. However, a small momentum value may not reliably avoid local minima. Conversely, the large value of momentum will cause the learning process (i.e., search) to oscillate and prevent convergence (Yu and Chen 1997). In short, large learning rate or small momentum may cause local minima issues. However, the choice of learning and momentum are dependent on the complexity of the data and the objective of using ANN. Our experiment showed that the low value of learning rates and relatively small momentums have better and reliable generalization performance. However, a learning rate between 0.2 and 0.5 and a momentum term between 0.4 and 0.5 seem to provide the appropriate combination of the pavement performance model (Attoh-Okine 1999). Further, learning rate plays a dominant role (compared to momentum) in our hyperparameter optimization process. In our case study, the low value of learning rates (0.01–0.04) with momentums less than 0.6 are more appropriate for training ANN-based land price evaluation.

4.6.2.1 Semivariogram-Based Spatial Sampling for Handling Inter-Sample Dependence

Spatial statistical approaches used in this study, represented by semivariogram and spatial interpolation, provide support for hyperparameter selection of neural network-based modeling. The effective sample size approach based on spatial autocorrelation proposed by Griffith (2005) allows for determining the appropriate sample size by considering (spatial) autocorrelation among sampled observations. Based on Griffith's (2005) effective sample size approach, we only needed 88 sampled hyperparameter sets (with consideration of spatial autocorrelation) instead of 383 suggested by conventional statistics (without taking into account dependence among hyperparameter samples). The difference between classic sampling strategy (statistics here) and semivariogram-based spatial sampling strategy is that spatial sampling determines a minimal sample size by considering spatial dependence or here inter-sampled dependence in the hyperparameter space. The benefits of semivariogram-based spatial sampling have been well acknowledged over the past few decades in soil science (McBratney and Webster 1983; Flatman and Yfantis 1984; Lark 2002). However, the use of spatial sampling in other fields remains limited. In this study, the semivariogram results from grid search, and random search suggested that dependence among samples exists in hyperparameter space. Yet, a traditional semivariogram supports 2-dimensional space. It can be used in high-dimensional space, i.e., the multivariate semivariogram (Pebesma 2004; Wackernagel 2013). We will investigate the capability of multivariate semivariogram in hyperparameter optimization in future research.

According to our findings, the minimal sample size for hyperparameter selection is substantially reduced when autocorrelation is considered. This decrease in sample size is significant because of the computationally challenging nature of hyperparameter selection. In the experiment reported in this study, each hyperparameter set will require about 3.2 h of computing time (average computing time for 100 repetitions). Thus, reduction in effective sample size (from 383 to 88 here) based on the use of spatial statistical approaches in this study can greatly decrease the computational demand for hyperparameter optimization. Furthermore, the variance of sampled hyperparameter sets in random search is lower than those from grid search. This finding is consistent with McBratney and Webster (1983) and Delmelle (2014), both of which suggested that random sampling is often preferred over systematic sampling. In particular, random sampling is more suitable than systematic sampling for the estimation of semivariogram results. Bergstra et al. (2011) and Bergstra and

Bengio (2012) also suggested that random search can find models with more accuracy and efficiency than grid search in terms of configuring the hyperparameters of the neural network.

4.6.2.2 Spatial Interpolation for Prediction of Generalization Performance

The estimation of points with unknown values or missing data is well documented (Creutin and Obled 1982), and spatial interpolation was usually involved (Lam 1983; Mitas and Mitasova 1999). Kriging-based spatial interpolation method helps us to generate the continuous surface on the hyperparameter space. Integrating spatial interpolation with hyperparameter optimization could convert discrete representation (from samples) into the continuous surface. That is, spatial interpolation can predict the continuous hyperparameter space based on discrete sampled hyperparameter sets. In the prediction map based on spatial interpolation (Fig. 4.12), the generalization performance for both grid search and random search tends to be lower when both learning rates and momentums become larger. In the meanwhile, we generated the continuous surface for robustness analysis using spatial interpolation. Our prediction results (see Fig. 4.13) demonstrated that the generalization performance of the hyperparameter optimization approach for grid search and random search tends to be more stable when learning rates and momentums are smaller. The prediction ability of spatial interpolation has been examined by a number of studies in different fields, such as estimating missing data in climate change or hydrology (Tabios and Salas 1985; Jeffrey et al. 2001), or mapping of soil properties (Robinson and Metternicht 2006). However, the use of spatial interpolation in hyperparameter optimization has not been fully investigated. In this study, our results here provide evidence for the feasibility of prediction of spatial interpolation methods in hyperparameter space. In addition, the results of spatial interpolation illustrated the spatial variability of the hyperparameters studied. This would enable the identification of direction for future selection of hyperparameters. The future selection is also known as second-phase sampling strategy, which takes additional samples to improve the overall prediction accuracy (Delmelle 2014). However, additional samples are usually collected from the maximum kriging variance area (Delmelle and Goovaerts 2009). Whereas, in hyperparameter optimization, additional samples should be selected from the space that has better generalization performance. Second-phase sampling will refine the optimal hyperparameters and further show the variation at small scale.

In general, the optimal hyperparameters and effective sample size are dependent on the data and structure of ANNs. Hence, for each case study, it is better to use the hyperparameter optimization framework that we proposed to find the suitable settings for ANNs. Meanwhile, all of these benefits brought by this framework will give us more flexibility and capacity in terms of applications of the ANN approach into the modeling of complex adaptive spatial systems.

4.6.3 Computing Performance

We employed a set of strategies to resolve the computational challenge of hyperparameter optimization. At computing level, the results demonstrated that high performance computing substantially improves computing performance. Speedup is 59 with 100 CPUs– i.e., the entire experiment of random search needs 6.93 h with 100 CPUs instead of 321.51 h (about 13.4 days) based on 1 CPU. A number of applications and studies used parallel computing infrastructure in their hyperparameter optimization, and demonstrated the capability of high performance computing in accelerating the selection process of hyperparameter optimization (Bergstra et al. 2015; Kotthoff et al. 2016). However, the computational challenge at model level can be further addressed by leveraging information at the model level (spatial statistical approaches here). Semivariogram-based spatial sampling was involved in this study to estimate the effective sample size, and a Monte Carlo approach was applied to find the optimal number of repetitions for each hyperparameter set. Based on our results, the estimated computing time for the entire experiment is decreased to 281.6 h (about 11.7 days) based on 1 CPU (about 1.7 days of reduction in computing time due to the introduction of spatial statistics approach). The optimal number of repetitions is less than 100 when the learning rate is less than 0.05 and the momentum is smaller than 0.7 using the approach suggested in this study.

4.7 Conclusion

In this study, we presented an automated hyperparameter optimization framework for the ANN-based spatial explicit modeling. Our approach demonstrated the utility of hyperparameter optimization in the GIScience domain and, in return, hyperparameter optimization could benefit from spatial statistical methods with respect to handling challenges facing hyperparameter optimization. To the best of our knowledge, this approach is the first in terms of using spatial statistical methods to facilitate and enhance hyperparameter optimization. Further, we showed that spatial interpolation algorithms could be applied to generate the continuous performance surface of hyperparameters. The neural networks selected through the two search methods outperform traditional linear regression modeling, and the generalization performance of the two search methods is consistent. Results in this study suggested that random search is more effective than grid search for the identification of optimal setting of hyperparameters associated neural network-based spatial modeling.

This hyperparameter optimization framework allows for integrating hyperparameter optimization and spatial statistical methods. The automated hyperparameter optimization approach addresses the limitations from traditional model configuration methods (e.g. trial and error) typically in the GIScience domain. Spatial statistical methods showed great potential in addressing the limitations of hyperparameter optimization (due to its independence assumption). Semivariogram-based spatial sampling strategy first revealed spatial dependence in hyperparameter space, then provided support for estimating effective sample size that considers spatial dependence. Furthermore, spatial interpolation methods showed their utility in predicting continuous patterns on a given hyperparameter space.

Because of higher cost and time-consuming nature of hyperparameter optimization, computational efficiency is the major challenge. Existing applications and our approach have demonstrated that high performance computing can address this challenge at computing level. More importantly, our study illustrated that spatial statistical approaches as a model-level solution could handle the computational bottleneck of hyperparameter optimization. For example, the spatial sampling strategy could reduce sample size (compared with conventional statistics) when spatial dependence exists, and semivariogram provides a suggestion for determining the suitable sampling methods.

Our future work will focus on a set of threads. First, we used two-dimensional (learning rate—momentum) hyperparameter space in this study. We will extend our approach into a higher-dimensional hyperparameter space such as three dimensions (learning rate—momentum—training iterations). Second, we will investigate the utility of second-phase sampling strategy in hyperparameter optimization. The second-phase sampling strategy aims to optimize the sampling design and investigate the spatial variability of hyperparameter optimization at small scale. Third, we will extend the automated selection framework in more use cases to examine the applicability of this hyperparameter optimization framework that takes into account spatial characteristics in hyperparameters. Forth, we will involve more hyperparameter optimization methods in our framework, such as genetic algorithm. Fifth, the vanishing gradient problem was found in training a neural network with gradient-based back-propagation learning algorithm. Although this problem greatly affects deep learning (the minimal number of hidden layers is greater than 3), vanishing gradient problem is a worthy considering problem in our future work.

References

Almeida, C.M., J.M. Gleriani, Emiliano Ferreira Castejon, and B.S. Soares-Filho. 2008. Using neural networks and cellular automata for modelling intra-urban land-use dynamics. *International Journal of Geographical Information Science* 22 (9): 943–963.

Andradóttir, Sigrún. 2006. An overview of simulation optimization via random search. *Handbooks in Operations Research and Management Science* 13: 617–631.

Anselin, Luc. 1995. Local indicators of spatial association—LISA. *Geographical Analysis* 27 (2): 93–115.

Anselin, Luc, Anil K. Bera, Raymond Florax, and Mann J. Yoon. 1996. Simple diagnostic tests for spatial dependence. *Regional Science and Urban Economics* 26 (1): 77–104.

Arribas, Iván., Fernando García, Francisco Guijarro, Javier Oliver, and Rima Tamošiūnienė. 2016. Mass appraisal of residential real estate using multilevel modelling. *International Journal of Strategic Property Management* 20 (1): 77–87.

Atack, Jeremy, and Robert A. Margo. 1998. "Location, location, location!" The price gradient for vacant urban land: New York, 1835 to 1900. *The Journal of Real Estate Finance and Economics* 16 (2): 151–172.

Atkins, Daniel. 2003. Revolutionizing science and engineering through cyberinfrastructure: Report of the National Science Foundation blue-ribbon advisory panel on cyberinfrastructure.

Attoh-Okine, Nii O. 1999. Analysis of learning rate and momentum term in backpropagation neural network algorithm trained to predict pavement performance. *Advances in Engineering Software* 30 (4):291–302.

Batista, Gustavo EAPA., and Maria Carolina Monard. 2003. An analysis of four missing data treatment methods for supervised learning. *Applied Artificial Intelligence* 17 (5–6): 519–533.

Berberoglu, S., Christopher D. Lloyd, P.M. Atkinson, and Paul J. Curran. 2000. The integration of spectral and textural information using neural networks for land cover mapping in the Mediterranean. *Computers & Geosciences* 26 (4): 385–396.

Bergstra, James, and Yoshua Bengio. 2012. Random search for hyper-parameter optimization. *Journal of Machine Learning Research* 13 (Feb):281–305.

Bergstra, James S, Rémi Bardenet, Yoshua Bengio, and Balázs Kégl. 2011. Algorithms for hyper-parameter optimization. *Advances in Neural Information Processing Systems.*

Bergstra, James, Dan Yamins, and David D Cox. 2013. Hyperopt: A python library for optimizing the hyperparameters of machine learning algorithms. In *Proceedings of the 12th Python in science conference.*

Bergstra, James, Brent Komer, Chris Eliasmith, Dan Yamins, and David D Cox. 2015. Hyperopt: a python library for model selection and hyperparameter optimization. *Computational Science & Discovery* 8(1):014008.

Bishop, Christopher M. 2006. *Pattern recognition and machine learning*. springer.

Biswajeet, Pradhan, and Lee Saro. 2007. Utilization of optical remote sensing data and GIS tools for regional landslide hazard analysis using an artificial neural network model. *Earth Science Frontiers* 14 (6): 143–151.

Bradshaw, Corey JA., Lloyd S. Davis, Martin Purvis, Qingqing Zhou, and George L. Benwell. 2002. Using artificial neural networks to model the suitability of coastline for breeding by New Zealand fur seals (*Arctocephalus forsteri*). *Ecological Modelling* 148 (2): 111–131.

Brigham, Eugene F. 1965. The determinants of residential land values. *Land Economics* 41 (4): 325–334.

Brown, Charles E. 1998. Coefficient of variation. In *Applied multivariate statistics in geohydrology and related sciences*, 155–157. Springer.

Chen, Sheng, S.A. Billings, and P.M. Grant. 1990. Non-linear system identification using neural networks. *International Journal of Control* 51 (6): 1191–1214.

Claesen, Marc, and Bart De Moor. 2015. Hyperparameter search in machine learning. arXiv preprint arXiv:1502.02127.

Creutin, JD, and Ch Obled. 1982. Objective analyses and mapping techniques for rainfall fields: An objective comparison. *Water resources research* 18(2):413–431.

Cunningham, Christopher R. 2006. House price uncertainty, timing of development, and vacant land prices: Evidence for real options in Seattle. *Journal of Urban Economics* 59 (1): 1–31.

Dai, Erfu, Shaohong Wu, Wenzhong Shi, Chui-kwan Cheung, and Ahmed Shaker. 2005. Modeling change-pattern-value dynamics on land use: an integrated GIS and artificial neural networks approach. *Environmental Management* 36(4):576–591.

Delmelle, Eric M., and Pierre Goovaerts. 2009. Second-phase sampling designs for non-stationary spatial variables. *Geoderma* 153 (1–2): 205–216.

Delmelle, Eric M. 2014. Spatial sampling. In *Handbook of regional science*, 1385–1399. Springer.

Demuth, Howard B, Mark H Beale, Orlando De Jess, and Martin T Hagan. 2014. *Neural network design*. Martin Hagan.

Dreiseitl, Stephan, and Lucila Ohno-Machado. 2002. Logistic regression and artificial neural network classification models: A methodology review. *Journal of biomedical informatics* 35(5):352–359.

Erbek, F Sunar, C Özkan, and M Taberner. 2004. Comparison of maximum likelihood classification method with supervised artificial neural network algorithms for land use activities. *International Journal of Remote Sensing* 25(9):1733–1748.

Quigley, John M. 2002. Real estate prices and economic cycles. *Berkeley Program on Housing and Urban Policy.*

Gopal, Sucharita. 2017. Artificial neural networks in geospatial analysis. *The International Encyclopedia of Geography.*

Fawcett, Tom. 2006. An introduction to ROC analysis. *Pattern Recognition Letters* 27 (8): 861–874.

Fischer, Manfred M, Martin Reismann, and Katerina Hlavackova–Schindler. 2003. Neural network modeling of constrained spatial interaction flows: Design, estimation, and performance issues. *Journal of Regional Science* 43 (1):35–61.

Flatman, George T, and Angelo A Yfantis. 1984. Geostatistical strategy for soil sampling: the survey and the census. *Environmental monitoring and assessment* 4 (4):335–349.

Fujita, Masahisa, Paul R Krugman, Anthony J Venables, and Massahisa Fujita. 1999. *The spatial economy: Cities, regions and international trade*, vol. 213. Wiley Online Library.

Girouard, Nathalie, and Sveinbjörn Blöndal. 2001. House prices and economic activity.

Godden, Bill. 2004. Sample size formulas. Retrieved on December 3:2013.

Goethals, Peter LM., Andy P. Dedecker, Wim Gabriels, Sovan Lek, and Niels De Pauw. 2007. Applications of artificial neural networks predicting macroinvertebrates in freshwaters. *Aquatic Ecology* 41 (3): 491–508.

Gopal, Sucharita, and Curtis Woodcock. 1996. Remote sensing of forest change using artificial neural networks. *IEEE Transactions on Geoscience and Remote Sensing* 34 (2): 398–404.

Govindaraju, Rao S, and Adiseshappa Ramachandra Rao. 2013. *Artificial neural networks in hydrology*, vol. 36. Springer Science & Business Media.

Grekousis, George, Panos Manetos, and Yorgos N Photis. 2013. Modeling urban evolution using neural networks, fuzzy logic and GIS: The case of the Athens metropolitan area. *Cities* 30:193–203.

Grekousis, George, and Yorgos N. Photis. 2014. Analyzing high-risk emergency areas with GIS and neural networks: The case of Athens, Greece. *The Professional Geographer* 66 (1): 124–137.

Griffith, Daniel A. 2005. Effective geographic sample size in the presence of spatial autocorrelation. *Annals of the Association of American Geographers* 95 (4): 740–760.

Hahnloser, Richard HR, Rahul Sarpeshkar, Misha A Mahowald, Rodney J Douglas, and H Sebastian Seung. 2000. Digital selection and analogue amplification coexist in a cortex-inspired silicon circuit. *Nature* 405(6789):947–951.

Handcock, Mark S., and James R. Wallis. 1994. An approach to statistical spatial-temporal modeling of meteorological fields. *Journal of the American Statistical Association* 89 (426): 368–378.

Heermann, Philip Dale, and Nahid Khazenie. 1992. Classification of multispectral remote sensing data using a back-propagation neural network. *IEEE Transactions on Geoscience and Remote Sensing* 30 (1): 81–88.

Heikkila, Eric, Peter Gordon, Jae Ik Kim, Richard B Peiser, Harry W Richardson, and David Dale-Johnson. 1989. What happened to the CBD-distance gradient? Land values in a policentric city. *Environment and planning A* 21(2):221–232.

Hornik, Kurt, Maxwell Stinchcombe, and Halbert White. 1989. Multilayer feedforward networks are universal approximators. *Neural Networks* 2 (5): 359–366.

Hsu, Kuo-lin, Hoshin Vijai Gupta, and Soroosh Sorooshian. 1995. Artificial neural network modeling of the rainfall-runoff process. *Water resources research* 31(10):2517–2530.

Hu, Shougeng, Shengfu Yang, Weidong Li, Chuanrong Zhang, and Feng Xu. 2016. Spatially non-stationary relationships between urban residential land price and impact factors in Wuhan city, China. *Applied Geography* 68:48–56.

Isik, Sabahattin, Latif Kalin, Jon E. Schoonover, Puneet Srivastava, B. Graeme, and Lockaby. 2013. Modeling effects of changing land use/cover on daily streamflow: An artificial neural network and curve number based hybrid approach. *Journal of Hydrology* 485: 103–112.

Jeffrey, Stephen J., John O. Carter, Keith B. Moodie, and Alan R. Beswick. 2001. Using spatial interpolation to construct a comprehensive archive of Australian climate data. *Environmental Modelling & Software* 16 (4): 309–330.

Joy, Michael K., and Russell G. Death. 2004. Predictive modelling and spatial mapping of freshwater fish and decapod assemblages using GIS and neural networks. *Freshwater Biology* 49 (8): 1036–1052.

Kalos, Malvin H, and Paula A Whitlock. 2008. *Monte carlo methods*. Wiley.

Karsoliya, Saurabh. 2012. Approximating number of hidden layer neurons in multiple hidden layer BPNN architecture. *International Journal of Engineering Trends and Technology* 3 (6): 714–717.

Kavzoglu, Taskin, and Paul M. Mather. 2003. The use of backpropagating artificial neural networks in land cover classification. *International Journal of Remote Sensing* 24 (23): 4907–4938.

Kia, Masoud Bakhtyari, Saied Pirasteh, Biswajeet Pradhan, Ahmad Rodzi Mahmud, Wan Nor Azmin Sulaiman, and Abbas Moradi. 2012. An artificial neural network model for flood simulation using GIS: Johor River Basin, Malaysia. *Environmental Earth Sciences* 67(1):251–264.

Kotsiantis, Sotiris B., I. Zaharakis, and P. Pintelas. 2007. Supervised machine learning: A review of classification techniques. *Emerging Artificial Intelligence Applications in Computer Engineering* 160: 3–24.

Kotthoff, Lars, Chris Thornton, Holger H. Hoos, Frank Hutter, and Kevin Leyton-Brown. 2016. Auto-WEKA 2.0: Automatic model selection and hyperparameter optimization in WEKA. *Journal of Machine Learning Research* 17: 1–5.

Krejcie, Robert V, and Daryle W Morgan. 1970. Determining sample size for research activities. *Educational and psychological measurement* 30(3):607–610.

Krige, Danie G. 1978. *Lognormal-de Wijsian geostatistics for ore evaluation*. South African Institute of mining and metallurgy Johannesburg.

Lam, Nina Siu-Ngan. 1983. Spatial interpolation methods: A review. *The American Cartographer* 10 (2): 129–150.

Lark, R.M. 2002. Optimized spatial sampling of soil for estimation of the variogram by maximum likelihood. *Geoderma* 105 (1–2): 49–80.

LaValle, Steven M., Michael S. Branicky, and Stephen R. Lindemann. 2004. On the relationship between classical grid search and probabilistic roadmaps. *The International Journal of Robotics Research* 23 (7–8): 673–692.

Legendre, Pierre, and Marie Josée Fortin. 1989. Spatial pattern and ecological analysis. *Vegetatio* 80(2):107–138.

Lerman, PM. 1980. Fitting segmented regression models by grid search. *Applied Statistics* 77–84.

Li, Xia, and Anthony Gar-On Yeh. 2002. Neural-network-based cellular automata for simulating multiple land use changes using GIS. *International Journal of Geographical Information Science* 16(4):323–343.

Li, Xiaodong, Feng Ling, Yun Du, Qi Feng, and Yihang Zhang. 2014. A spatial–temporal Hopfield neural network approach for super-resolution land cover mapping with multi-temporal different resolution remotely sensed images. *ISPRS Journal of Photogrammetry and Remote Sensing* 93:76–87.

Linares-Rodriguez, Alvaro, José Antonio Ruiz-Arias, David Pozo-Vazquez, and Joaquin Tovar-Pescador. 2013. An artificial neural network ensemble model for estimating global solar radiation from Meteosat satellite images. *Energy* 61:636–645.

Maa, C.-Y., and M.A. Schanblatt. 1992. A two-phase optimization neural network. *IEEE Transactions on Neural Networks* 3 (6): 1003–1009.

Mas, Jean F., and Juan J. Flores. 2008. The application of artificial neural networks to the analysis of remotely sensed data. *International Journal of Remote Sensing* 29 (3): 617–663.

Mas, Jean-François., Henri Puig, José Luis. Palacio, and Atahualpa Sosa-López. 2004. Modelling deforestation using GIS and artificial neural networks. *Environmental Modelling & Software* 19 (5): 461–471.

McBratney, A.B., and R. Webster. 1983. HOW Many observations are needed for regional estimation of soil properties? *Soil Science* 135 (3): 177–183.

McDonald, John H. 2009. *Handbook of biological statistics*, vol. 2. Sparky House Publishing Baltimore, MD.

McKay, Michael D., Richard J. Beckman, and William J. Conover. 1979. Comparison of three methods for selecting values of input variables in the analysis of output from a computer code. *Technometrics* 21 (2): 239–245.

Mera, Koichi, and Bertrand Renaud. 2016. *Asia's financial crisis and the role of real estate*. Routledge.

Miller, Diane M., Edit J. Kaminsky, and Soraya Rana. 1995. Neural network classification of remote-sensing data. *Computers & Geosciences* 21 (3): 377–386.

Mitas, Lubos, and Helena Mitasova. 1999. Spatial interpolation. *Geographical Information Systems: Principles, Techniques, Management and Applications* 1: 481–492.

Moran, Patrick AP.. 1950. Notes on continuous stochastic phenomena. *Biometrika* 37 (1/2): 17–23.

Nayak, Purna C, YR Satyaji Rao, and KP Sudheer. 2006. Groundwater level forecasting in a shallow aquifer using artificial neural network approach. *Water Resources Management* 20(1):77–90.

Nevtipilova, Veronika, Justyna Pastwa, Mukesh Singh Boori, and Vit Vozenilek. 2014. Testing artificial neural network (ANN) for spatial interpolation. *International Journal of Geology and Geosciences (JGG), ISSN 2329* 6755:01–09.

Nourani, Vahid, Aida Hosseini Baghanam, Jan Adamowski, and Mekonnen Gebremichael. 2013. Using self-organizing maps and wavelet transforms for space–time pre-processing of satellite precipitation and runoff data in neural network based rainfall–runoff modeling. *Journal of Hydrology* 476: 228–243.

Olden, Julian D., and Donald A. Jackson. 2002. Illuminating the "black box": A randomization approach for understanding variable contributions in artificial neural networks. *Ecological Modelling* 154 (1): 135–150.

Openshaw, Stan, and Christine Openshaw. 1997. *Artificial intelligence in geography*. Wiley.

Özesmi, Stacy L., and Uygar Özesmi. 1999. An artificial neural network approach to spatial habitat modelling with interspecific interaction. *Ecological Modelling* 116 (1): 15–31.

Paola, J.D., and R.A. Schowengerdt. 1995. A review and analysis of backpropagation neural networks for classification of remotely-sensed multi-spectral imagery. *International Journal of Remote Sensing* 16 (16): 3033–3058.

Pebesma, Edzer J. 2004. Multivariable geostatistics in S: the gstat package. *Computers & Geosciences* 30(7):683–691

Pijanowski, Bryan C, Daniel G Brown, Bradley A Shellito, and Gaurav A Manik. 2002. Using neural networks and GIS to forecast land use changes: a land transformation model. *Computers, environment and urban systems* 26(6):553–575

Pijanowski, Bryan C, Snehal Pithadia, Bradley A Shellito, and Konstantinos Alexandridis. 2005. Calibrating a neural network-based urban change model for two metropolitan areas of the Upper Midwest of the United States. *International Journal of Geographical Information Science* 19(2):197–215

Pijanowski, Bryan C, Amin Tayyebi, Jarrod Doucette, Burak K Pekin, David Braun, and James Plourde. 2014. A big data urban growth simulation at a national scale: Configuring the GIS and neural network based Land Transformation Model to run in a High Performance Computing (HPC) environment. *Environmental Modelling & Software* 51:250–268

Pradhan, Biswajeet, and Saro Lee. 2010. Landslide susceptibility assessment and factor effect analysis: Backpropagation artificial neural networks and their comparison with frequency ratio and bivariate logistic regression modelling. *Environmental Modelling & Software* 25 (6): 747–759.

Quan, Daniel C., and Sheridan Titman. 1999. Do real estate prices and stock prices move together? An international analysis. *Real Estate Economics* 27 (2): 183–207.

Rigol, Juan P., Claire H. Jarvis, and Neil Stuart. 2001. Artificial neural networks as a tool for spatial interpolation. *International Journal of Geographical Information Science* 15 (4): 323–343.

Rigol-Sanchez, JP, M Chica-Olmo, and F Abarca-Hernandez. 2003. Artificial neural networks as a tool for mineral potential mapping with GIS. *International Journal of Remote Sensing* 24(5):1151–1156.

Robinson, T.P., and G. Metternicht. 2006. Testing the performance of spatial interpolation techniques for mapping soil properties. *Computers and Electronics in Agriculture* 50 (2): 97–108.

Rumelhart, David E, James L McClelland, and PDP Research Group. 1988. *Parallel distributed processing*. Vol. 1: IEEE.

Särndal, Carl-Erik, Ib Thomsen, Jan M Hoem, DV Lindley, O Barndorff-Nielsen, and Tore Dalenius. 1978. Design-based and model-based inference in survey sampling [with discussion and reply]. *Scandinavian Journal of Statistics* 27–52.

Specht, Donald F. 1990. Probabilistic neural networks. *Neural Networks* 3 (1): 109–118.

Stathakis, D. 2009. How many hidden layers and nodes? *International Journal of Remote Sensing* 30 (8): 2133–2147.

Tabios, Guillermo Q., and Jose D. Salas. 1985. A comparative analysis of techniques for spatial interpolation of precipitation. *JAWRA Journal of the American Water Resources Association* 21 (3): 365–380.

Tang, Wenwu, and Meijuan Jia. 2014. Global sensitivity analysis of a large agent-based model of spatial opinion exchange: A heterogeneous multi-GPU acceleration approach. *Annals of the Association of American Geographers* 104 (3): 485–509.

Tang, Wenwu, George P. Malanson, and Barbara Entwisle. 2009. Simulated village locations in Thailand: A multi-scale model including a neural network approach. *Landscape Ecology* 24 (4): 557–575.

Thornton, Chris, Frank Hutter, Holger H Hoos, and Kevin Leyton-Brown. 2013. Auto-WEKA: Combined selection and hyperparameter optimization of classification algorithms. In *Proceedings of the 19th ACM SIGKDD international conference on knowledge discovery and data mining*.

Tilman, David, and Peter M Kareiva. 1997. *Spatial ecology: the role of space in population dynamics and interspecific interactions*, vol. 30. Princeton University Press.

Tobler, Waldo R. 1970. A computer movie simulating urban growth in the Detroit region. *Economic Geography* 46 (sup1): 234–240.

Wackernagel, Hans. 2013. *Multivariate geostatistics: an introduction with applications*. Springer Science & Business Media.

Wilkinson, Barry, and Michael Allen. 1999. *Parallel programming*, vol. 999. Prentice hall Upper Saddle River, NJ.

Willmott, Cort J. 1982. Some comments on the evaluation of model performance. *Bulletin of the American Meteorological Society* 63(11):1309–1313.

Yamazaki, Ritsuko. 2001. Empirical testing of real option pricing models using Land Price Index in Japan. *Journal of Property Investment & Finance* 19 (1): 53–72.

Yu, Xiao-Hu, and Guo-An Chen. 1997. Efficient backpropagation learning using optimal learning rate and momentum. *Neural Networks* 10(3):517–527.

Zimmerman, Dale, Claire Pavlik, Amy Ruggles, and Marc P. Armstrong. 1999. An experimental comparison of ordinary and universal kriging and inverse distance weighting. *Mathematical Geology* 31 (4): 375–390.

Chapter 5
Study II. Spatially Explicit Hyperparameter Optimization of Neural Networks Accelerated Using High-Performance Computing

The major purpose of this chapter is to demonstrate the performance of the automated spatially explicit hyperparameter optimization (corresponding to objective 2). More specifically, I use the same case study area and datasets as Chap. 4 to illustrate the proposed spatially explicit hyperparameter optimization approach.

5.1 Introduction

The power of neural networks has been proved by a number of studies, but the selection of hyperparameters of neural networks (i.e., parameter settings) is still a "black or grey box" for a series of research fields, such as geography. Since the last century, a number of researchers suggested that good results depend on appropriate hyperparameters of neural networks and problem-specific parameter settings. Past experiences of hyperparameter optimization demonstrated that the searching surface of hyperparameters is "vast, undifferentiable, epistatic, complex, noisy, deceptive, multimodal surface" (P381) (Miller et al. 1989). And, the number of possible hyper-parameter sets can be unbounded, and changes in those hyperparameter sets should have a discontinuous effect in order to achieve a stable and accurate result (Miller et al. 1989; Weiß 1994; Branke 1995). Moreover, these researchers mentioned that a global hyperparameter optimization approach is needed for examining hyperpa-rameters of neural networks, such as evolutionary algorithm-based hyperparameter optimization (Branke 1995).

EAs have four main operators, including initialization, representation, selection, and variation operators. Only individuals get a competitive possibility, which is based on their fitness values, to survive long enough to produce next generation. Also, EAs reduce the risk of converging to local optimum because EAs search multiple regions (different chromosomes here) in the search space simultaneously and use the probabilistic rules to guide the search (Branke 1995). This searching process, in

M. Zheng, *Spatially Explicit Hyperparameter Optimization for Neural Networks*,
https://doi.org/10.1007/978-981-16-5399-5_5

turn, can be seen as a topology surface defined by trained network performance with related hyperparameters (Miller et al. 1989). For more details of EA, see Sect. 2.4.

A number of researchers already applied EAs to the hyperparameter optimization study for neural networks. Jin et al. (2004) applied EAs to minimize the approximation error and find the appropriate regularization terms. They used the scalarization approach to transform bi-objective to one objective. Meanwhile, they take advantage of another feature of EAs—their ability to run multiple models at the same time. Recently, some studies optimize deep learning hyperparameters through EAs. Multi-node Evolutionary Neural Networks for Deep Learning (MENNDL) was proposed to automatically select optimal hyperparameters of convolutional neural networks based on GAs (Young et al. 2015). In order to address the computing challenges, they implemented MENDL in the high-performance and parallel computing platform. But, they found that EAs are unable to maximize the use of available computing resources. In other words, the computing time for models with different hyperparameters is significantly varying. Some processors might finish earlier and stay idle.

As Zheng et al. (2019) discussed, hyperparameter optimization with consideration of spatial dependence could improve model- and computing-level computing performance. Also, methods from GIScience (spatial statistics) could explore the landscape of hyperparameter space, and generate generalization performance based on sampled hyperparameters. However, there are several limitations in their study. First, their hyperparameter optimization approach is a manual approach. The sampled hyperparameters are manually selected through sampling methods. Second, although they discussed second-phase sampling strategy is important and necessary to find the accurate and stable results, they did not incorporate second-phase sampling strategy into their framework. Third, the ability of exploration and exploitation is weak in their framework. Thus, an automated and comprehensive hyperparameter optimization framework with consideration spatial dependence is necessary for addressing these limitations.

The rest of this study is organized in the following manner. Section 5.2 discusses study area and related datasets, the methodology of spatially explicit hyperparameter optimization is presented in Sect. 5.3. Sections 5.4, 5.5 and 5.6 present implementation, experimental results and discussions. The conclusion is shown in Sect. 5.7.

5.2 Study Area and Data

The study area in this case study is Mecklenburg County, North Carolina, USA (see Fig. 5.1). The largest city in Mecklenburg County is Charlotte with 905,318 population in 2020, whose population ranks 15th in the USA. Also, Charlotte is one of the fastest-growing cities, around 60 people moving to Charlotte every day. The median housing price (sales price) in April 2020 was $290,380, which has

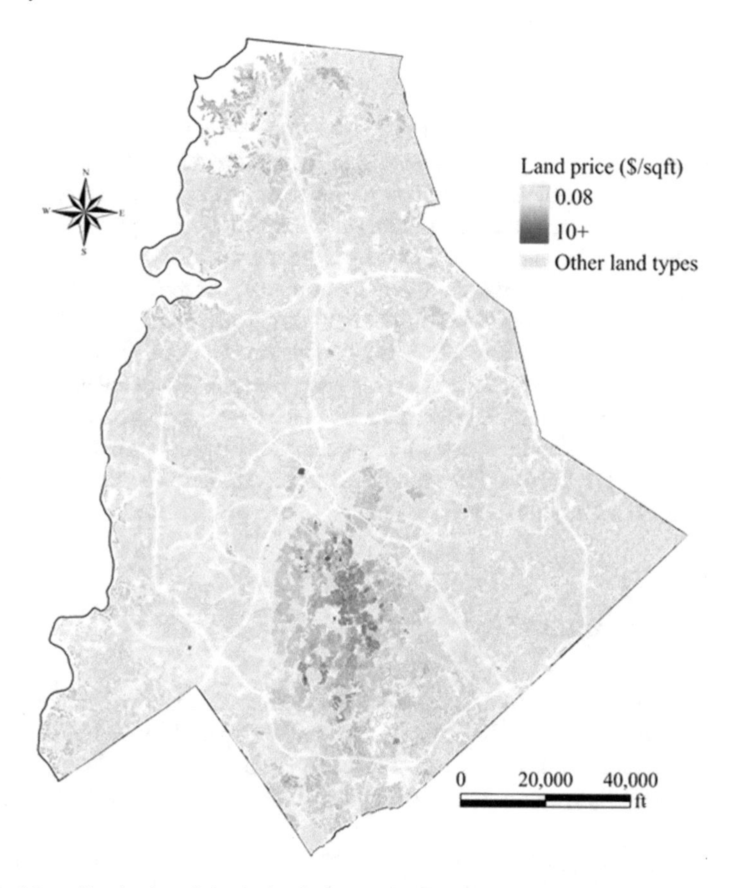

Fig. 5.1 Map of land price of single-family houses in Charlotte, North Carolina

11.7% year-over-year change. Over 4,500 housings (including single-family houses, townhouses, condos/apartments) were sold between January to April, 2020.

Followed Zheng et al.'s study (2019), in this study, I adopt six driving factors to investigate the relationship between social environmental variables and land price of single-family houses. All datasets were obtained from Data Center of Mecklenburg County (http://maps.co.mecklenburg.nc.us/openmapping), including land price, location of uptown Charlotte, locations of parks (includes recreation parks and greenways), locations of public and private schools (includes elementary, middle and high schools), locations of major roads, and locations of hospitals. All distance-based variables are Euclidean distance. I selected a part of records as the experimental dataset, please refer to Zheng et al. (2019) for more information. The experimental dataset has 9,254 records, I further split them into a training dataset (6,478 records; 70%) and a testing dataset (2,776; 30%).

5.3 Methodology

The framework covers three modules, generations of sampled hyperparameters, evaluation of sampled hyperparameters, and hyperparameter analysis. In order to implement and automate spatially explicit hyperparameter optimization approach, three components (e.g., automatic search of hyperparameters, spatial prediction of hyperparameter space, and acceleration of hyperparameter search) involve in the framework. EA (automatic search of hyperparameters component) is used to automate the framework from generation of sampled hyperparameter to evaluation of sampled hyperparameters. Spatial interpolation methods (spatial prediction of hyperparameter space component) focus on examining the spatial dependence of sampled hyperparameters and generating a continuous space based on those discrete sampled hyperparameters. HPC is in charge of accelerating the hyperparameter search from computing-level. For detailed information about this framework, please see Chap. 3 and Zheng et al.'s study (2019).

5.4 Implementation

Learning rate and momentum are examined in this case study. Moreover, referring to other cases of applying EAs in the field of spatial problems (Xiao et al. 2002; Bennett et al. 2004; Cao et al. 2011, 2014), the total size of population, the number of generations, crossover rate, and mutation are set to 100, 100, 0.8, and 0.01, respectively.

I apply the automated spatially explicit hyperparameter optimization approach in a Linux computing cluster (Copperhead) from University Research Computing at the University of North Carolina at Charlotte (https://urc.uncc.edu/). Copperhead cluster has 88 nodes with CPU (1980 computing cores) connected through an infiniband interconnect network and eight nodes with GPU (80 computing cores). The spatially explicit hyperparameter optimization approach is written in Python scripting language. The libraries that we used in this study are Keras (https://keras.io/), TensorFlow (https://www.tensorflow.org/), PyKrige (https://pypi.org/project/PyKrige/), and pandas (https://pandas.pydata.org/).

5.5 Results

5.5.1 *Model Performance*

Testing MSEs was used to evaluate the performance of spatially explicit hyperparameter optimization. The maximum number of generations in this study is set to 100. Figure 5.2 illustrates the learning curve and distribution of testing MSEs (gener-

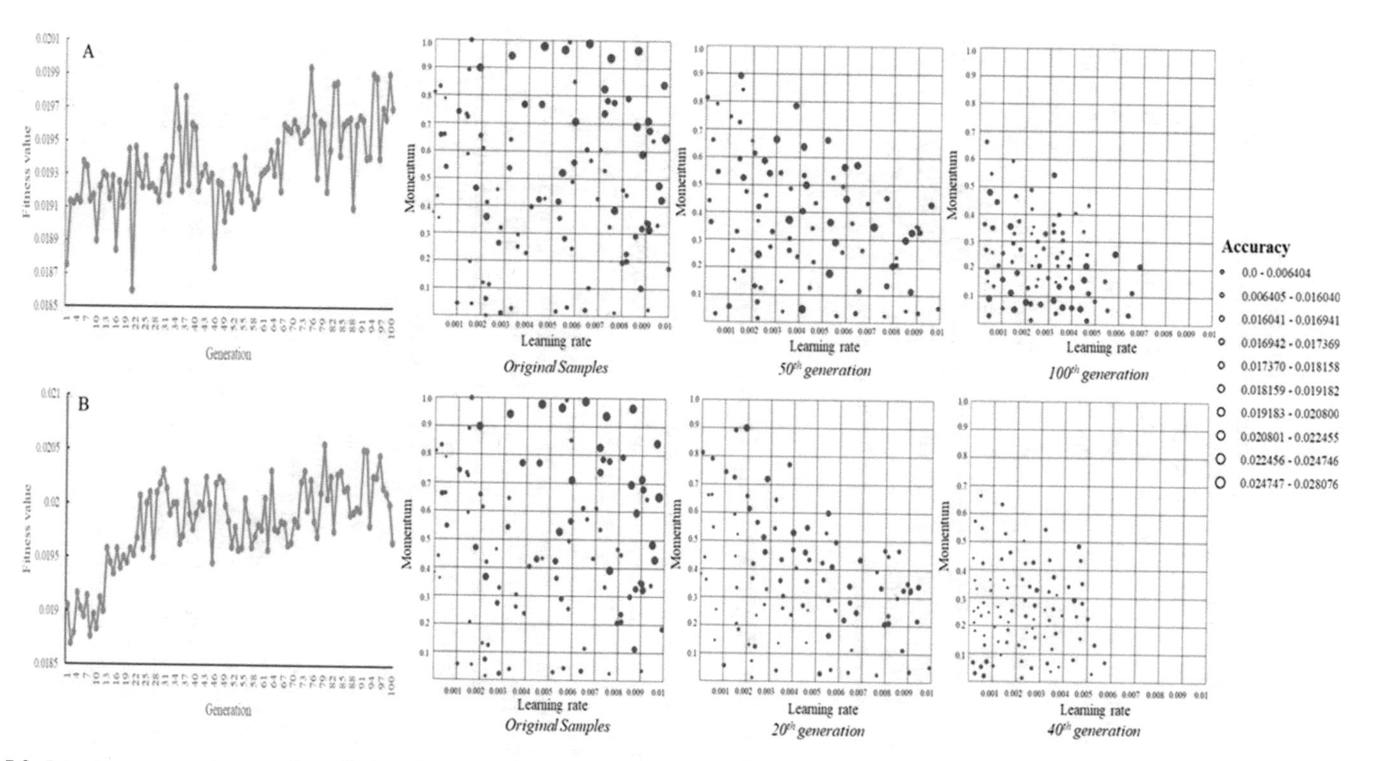

Fig. 5.2 Learning curve and scatterplots of MSEs for hyperparameter sets over different generations (**a** red circle is sampled points that generated by conventional EA-based hyperparameter optimization; **b** blue circle is sampled points that created by spatially explicit hyperparameter optimization; circle size from small to large stands for low MSE to large MSE; outliers were excluded)

alization performance) for the conventional EA-based hyperparameter optimization and our proposed approach. These scatterplots show the distributions of the original sampled points, the middle generation of the evolutional process, and the last generation that has the convergent results (i.e., the error difference among five generations is less than 0.1%). The ranges of learning rate and momentum in this case study are from 0 to 0.01 and from 0 to 1, respectively.

From the results of learning curves, we could observe that my approach can get the converged result more quickly than the conventional approach. In other words, the learning curve of the conventional approach has a constant pattern after the 80th generation (Fig. 5.2a), whereas my approach only needs around 30 generations (Fig. 5.2b). The scatterplots of MSEs visualize the training process. For those two approaches, high testing errors concentrate around the upper right and right regions during the entire process. The highest testing errors were caused by relatively larger learning rates (above 0.005) and larger momentum (above 0.7). In general, smaller values of learning rate and momentum cause a high generalization performance of neural networks. In the meanwhile, spatially explicit hyperparameter optimization approach uses less computing time to get the final results.

5.5.2 *Prediction Performance of Hyperparameters*

Figure 5.3 shows the prediction maps for conventional EA-based hyperparameter optimization approach and spatially explicit hyperparameter optimization approach. In the processes of EA-based hyperparameter optimization approaches, larger predicted errors concentrate on the upper right and upper right corner regions, particularly when momentum is greater than 0.8. The patterns in the middle region (i.e., learning rate is from 0.003 to 0.007 and momentum is from 0.3 to 0.6) of those prediction maps were continuously changed. The lower prediction errors occur in the leftmost region (learning rate is less than 0.002).

The prediction maps of the conventional EA-based approach is shown in Fig. 5.3a (RMSE[1]s of cross-validation is 2.35×10^{-3}). The increasing learning rates along with larger predicted errors. A higher prediction error occurs when momentum is larger than 0.8. We can obtain lower predicted errors with small learning rate (<0.002) and momentum (<0.5). For the prediction maps using spatially explicit hyperparameter optimization approach (Fig. 5.3b), RMSEs of cross-validation is 1.62×10^{-3}. In other words, the predicted error will slightly increase with increments in learning rates. Larger momentum (above 0.7) causes a higher predicted error. In addition, these two approaches obtain different results in the middle region (learning rate is from 0.003 to 0.007 and momentum is from 0.3 to 0.6). The pattern of the conventional approach has a smooth pattern, but my approach shows there are some hummocks and hollows. One of the reasons is that the prediction map of my approach is updated

[1] Root mean squared error (RMSE) is the square root of mean squared error (MSE).

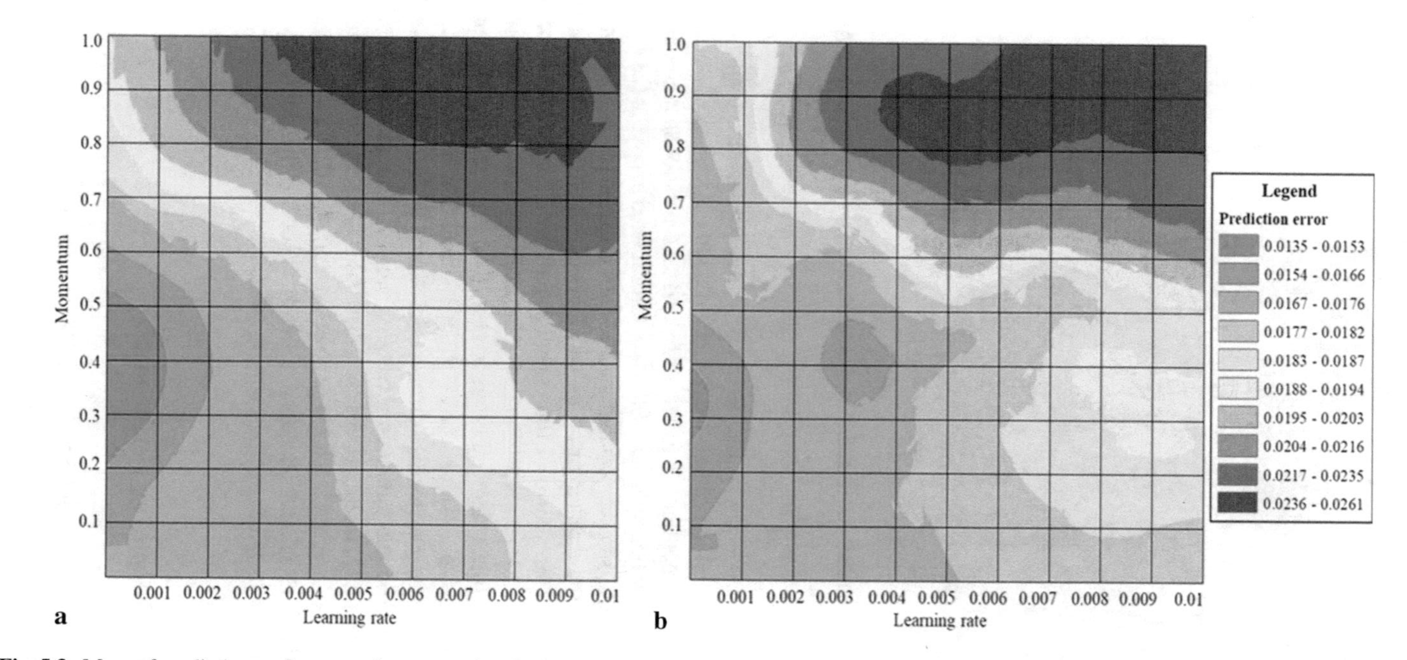

Fig. 5.3 Maps of prediction performance for conventional EA-based hyperparameter optimization and spatially explicit hyperparameter optimization (**a** conventional EA-based hyperparameter optimization using the 100th generation result; **b** spatially explicit hyperparameter optimization; prediction error: mean squared error)

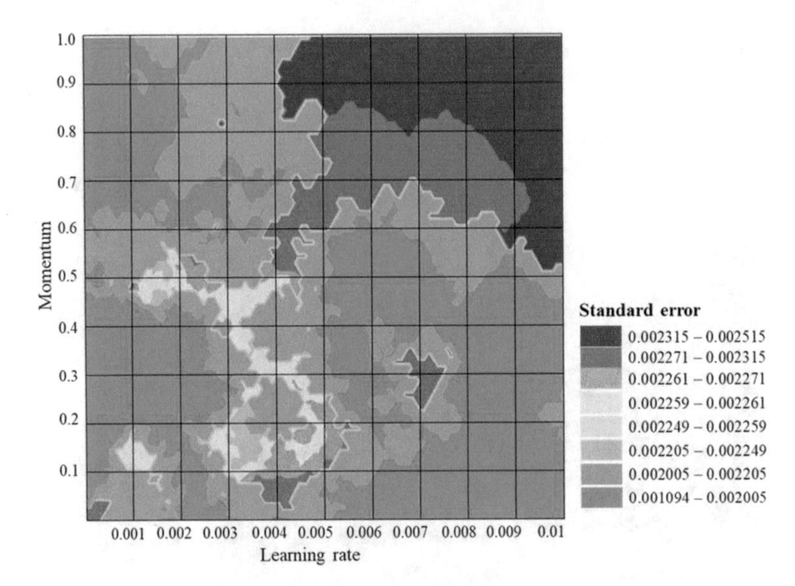

Fig. 5.4 Distribution of standard error based on testing MSEs using spatially explicit hyperparameter optimization approach

each generation by considering spatial dependence, and all sampled points during the entire process are used to generate the final prediction map.

Figure 5.4 shows the continuous patterns of standard errors for spatially explicit hyperparameter optimization approach. Standard error is used to refer to the standard deviation of a statistical sample population. Standard error serves as a measure of variation, the smaller the standard error, the result is more accurate, and the sample population is more representative of the overall population. The RMSE for cross-validation for the standard error is 3.16×10^{-3}. The results indicate that my approach is robust and accurate (most standard errors are less than 2.5×10^{-3}). The generalization performance is high when small values of learning rate and momentum are used. Specifically, the largest standard errors concentrated in the area with learning rate between 0.005–0.01 and momentum between 0.6–1.0. The most stable results are located in two areas: (1) learning rate is less than 0.003 and momentum within the range of 0.2–0.5 and 0.7–1, or (2) learning rate between 0.006–0.01 and momentum is less than 0.5. However, relatively larger standard errors occur when relatively large values of learning rate (>0.005) and momentum (>0.6).

5.5.3 Parallel Computing Performance

In this study, I assigned the entire task into 100 sub-tasks because there are 100 sampled hyperparameter sets in each generation. That is, each CPU works on a single sampled hyperparameter set. The estimated computing time (1 CPU used) of spatially explicit hyperparameter optimization approach was around 1,169,280 s (about 324.8 h) when 100,000 sampled hyperparameter sets involved. When I applied spatially explicit hyperparameter optimization on the HPC platform (Linux-based computing cluster), the computing time decreased to 77,868 s (about 21.6 h). The speedup was around 15. From Fig. 5.2, the proposed EA-based hyperparameter optimization can further reduce the computing time because it uses fewer generations to get the convergent results. By considering the convergency (converged at 40th generation), the hyperparameter optimization approach proposed in this study, only needed 31,104 s (estimated value; about 8.64 h). The speedup increased to 34.38 when our approach was adopted.

5.6 Discussions

In Zheng et al.'s study, they already demonstrated the ability of hyperparameter optimization by taking into account spatial dependence (Zheng et al. 2019). In this study, an automated and comprehensive framework of spatially explicit hyperparameter optimization approach is proposed, which considers spatial dependence, landscape of hyperparameter space, and multiple optimization results at the same time.

5.6.1 The Prediction of Generalization Performance

In the prediction maps of generalization performance (see Fig. 5.3), both the conventional approach and spatially explicit approach have a better generalization performance when learning rates and momentums are small. As we know, learning rate controls how much to change the model in response to the estimated error when the model weights are updated. In this case study, the range of learning rate is set to 0 to 0.01, which are relatively small values for learning rates in most neural network-based spatial models. Although learning rate does not show the prominent influence in this case study, smaller learning rates have better generalization performance. Momentum can accelerate the training process and avoid local minima issue (Yu and Chen 1997). The result of the influence of momentum is that relatively smaller momentums and learning rates have better and stable generalization performance.

Furthermore, in order to check the uncertainty of the generalization performance, the continuous surface was generated based on standard errors of the generalization performance. Standard error is measuring the variability of the sampled population. The higher the value, the result has lower accuracy because it is not a good estimate of the population parameter. The prediction results of standard errors (Fig. 5.4) show that the generalization performance of my approach tends to be more stable and accurate with small values of learning rates and momentum. One of the reasons is that the sampled hyperparameter sets with lower MSEs may have a larger chance to keep in the EA process, and more sampled hyperparameters that are located in the lower MSE areas are generated during the EA process.

Standard second-phase sampling strategy usually collects additional samples from low accuracy areas (Delmelle 2014), which will increase computing time and obtain inappropriate results (Delmelle and Goovaerts 2009). The hyperparameter optimization approach proposed in this book not only add the second-phase sample strategy into the framework, but also collect additional samples from the area that has better generalization performance. Therefore, my hyperparameter optimization approach has a better model-level performance than the conventional EA-based hyperparameter optimization approach.

5.6.2 *Computing Performance*

Although a series of studies used EA-based hyperparameter optimization approach to find the suitable hyperparameters for machine learning algorithms, such as artificial neural networks (Leung et al. 2003), support vector machines (Wu et al. 2007), and deep learning techniques (Young et al. 2015), there is no study that incorporates the landscape pattern of the search space during the searching process. From Fig. 5.2, we can see that when the landscape pattern of the search space involved, the speed of convergence is substantially improved. The average computing time for conventional EA-based approach with 100 population size (i.e., 10,000 sampled hyperparameters, including duplicate sampled hyperparameters) and 100 generations need 20.3 h. However, my EA-based approach only needs around 8 h to complete the task and get similar results as the conventional EA-based approach do. From the computing results of Case Study I, the estimated parallel computing time is 281.6 h even though they use effective sample size and high performance computing (i.e., they ran the 88 sampled hyperparameters 100 times using 88 CPUs).

Moreover, Xiao et al. (2002) mentioned that prior knowledge can be used to design the crossover and mutation, and then convergence would be expedited. In this study, my results demonstrated that prior knowledge (the results from spatial interpolation) can expedite convergence and reduce the computing time (Fig. 5.2). These findings are consistent with Xiao et al.'s study.

5.7 Conclusion

In this study, I presented an automated spatially explicit hyperparameter optimization framework, which covers three components: automatic search of hyperparameters, spatial prediction of hyperparameter space, and acceleration of hyperparameter search. This framework aims to explore the landscape of hyperparameter space and further accelerate the computing performance at model- and computing-level. The results of generalization performance demonstrated that the framework could help in exploring and exploiting the landscape of hyperparameter space, and provide accurate and stable results. Also, from the learning curves, my approach achieved a similar result as a conventional EA-based hyperparameter optimization approach with less computing time, which showed that my approach can further reduce computing time.

Future works will focus on the following three aspects. First, I will add more hyperparameters in this framework. Second, I will extend the current framework to other machine learning algorithms, such as random forests or convolutional neural networks. Third, I will compare the performance of the current framework with the latest hyperparameter optimization approaches, such as Population Based Training for neural networks, and further improve the performance of this framework.

References

Bennett, David A., Ningchuan Xiao, and Marc P. Armstrong. 2004. Exploring the geographic consequences of public policies using evolutionary algorithms. *Annals of the Association of American Geographers* 94 (4): 827–847.

Branke, Jürgen. 1995. Evolutionary algorithms for neural network design and training.

Cao, Kai, Michael Batty, Bo Huang, Yan Liu, Le Yu, and Jiongfeng Chen. 2011. Spatial multi-objective land use optimization: Extensions to the non-dominated sorting genetic algorithm-II. *International Journal of Geographical Information Science* 25 (12): 1949–1969.

Cao, Kai, Bo. Huang, Manchun Li, and Wenwen Li. 2014. Calibrating a cellular automata model for understanding rural–urban land conversion: A Pareto front-based multi-objective optimization approach. *International Journal of Geographical Information Science* 28 (5): 1028–1046.

Delmelle, Eric M., and Pierre Goovaerts. 2009. Second-phase sampling designs for non-stationary spatial variables. *Geoderma* 153 (1–2): 205–216.

Delmelle, Eric M. 2014. Spatial sampling. In *Handbook of regional science*, 1385–1399. Springer.

Jin, Yaochu, Tatsuya Okabe, and Bernhard Sendhoff. 2004. Neural network regularization and ensembling using multi-objective evolutionary algorithms. In *Proceedings of the 2004 congress on evolutionary computation (IEEE Cat. No. 04TH8753).*

Leung, Frank Hung-Fat, Hak-Keung Lam, Sai-Ho Ling, and Peter Kwong-Shun Tam. 2003. Tuning of the structure and parameters of a neural network using an improved genetic algorithm. *IEEE Transactions on Neural Networks* 14 (1): 79–88.

Miller, Geoffrey F, Peter M. Todd, and Shailesh U. Hegde. 1989. *Designing neural networks using genetic algorithms.* ICGA.

Weiß, Gerhard. 1994. Neural networks and evolutionary computation. I. Hybrid approaches in artificial intelligence. In *Proceedings of the first IEEE conference on evolutionary computation.* IEEE World Congress on Computational Intelligence.

Wu, Chih-Hung, Gwo-Hshiung Tzeng, Yeong-Jia Goo, and Wen-Chang Fang. 2007. A real-valued genetic algorithm to optimize the parameters of support vector machine for predicting bankruptcy. *Expert Systems with Applications* 32 (2): 397–408.

Xiao, Ningchuan, David A. Bennett, and Marc P. Armstrong. 2002. Using evolutionary algorithms to generate alternatives for multiobjective site-search problems. *Environment and Planning A* 34 (4): 639–656.

Young, Steven R., Derek C. Rose, Thomas P. Karnowski, Seung-Hwan Lim, and Robert M. Patton. 2015. Optimizing deep learning hyper-parameters through an evolutionary algorithm. In *Proceedings of the workshop on machine learning in high-performance computing environments.*

Yu, Xiao-Hu, and Guo-An Chen. 1997. Efficient backpropagation learning using optimal learning rate and momentum. *Neural Networks* 10 (3): 517–527.

Zheng, Minrui, Wenwu Tang, and Xiang Zhao. 2019. Hyperparameter optimization of neural network-driven spatial models accelerated using cyber-enabled high-performance computing. *International Journal of Geographical Information Science*. https://doi.org/10.1080/13658816.2018.1530355.

Chapter 6
Study III. An Integration of Spatially Explicit Hyperparameter Optimization with Convolutional Neural Networks-Based Spatial Models

In this chapter, the study shows the performance of an integration of spatially explicit hyperparameter optimization approach with other machine learning algorithms (links to objective 3). That is, I adopt a land change model with an integration of neural networks and cellular automata.

6.1 Introduction

The factors of dynamic land changes are important contributors to understand and interpret changes of natural systems (e.g., watershed ecosystem) and human systems (Tang and Yang 2020). With the increasing population in the world, land cover and land changes becomes a popular topic. Urbanization is one of the hot topics in land cover and land change domain. According to the 2018 Revision of World Urbanization Prospects from United Nations Department of Economic and Social Affairs (UNDESA), 55% of the total population lived in urban areas. In 2050, the number is expected to increase to 68% (UNDESA 2018). The process of urbanization causes land changes, and leads to urban land expansion and threaten natural systems (Wang and Zhang 2001). A review paper mentioned climate changes, land use and land cover changes, and human activities are closely linked to changes in wetland landscape patterns (Bai et al. 2005).

A number of land change models have been developed to handle the nonstationary process and gain insight into human–environment interactions. Cellular automata (CA) and its related extensions (e.g., CA-Markov models, CA-based future land use simulation models, hybrid CA and neural network models) (Li and Yeh 2002; Liang et al. 2018; Li and Reynolds 1997), and agent-based models (ABMs) are the most commonly used approaches in landscape change simulation. Both CA and ABMs are bottom-up simulation approaches, consider the neighborhood information and use the demand-allocation framework (Batty and Xie 1994; Castle and Crooks 2006). However, the size of neighborhood in CA is fixed, while the size of neighborhood

M. Zheng, *Spatially Explicit Hyperparameter Optimization for Neural Networks*,
https://doi.org/10.1007/978-981-16-5399-5_6

in ABMs can be changed vary on time because agents are free to move and interact with others or with their environments. Furthermore, the structure of CA is simple and flexible, but it usually only follows a few rules to update the state, whereas ABM allows us to simulate the complex and uncertain landscape changes and provide more realistic simulations.

Although a series of studies of CA and ABMs revealed that we could simulate complex geographic processes based on several simple rules and a few spatial variables (Li and Yeh 2000; He et al. 2018), the complicated geographic process cannot examine thoroughly by using CA and ABMs. With the development of computer technology, more and more studies use machine learning algorithms (e.g., artificial neural networks, random forests, genetic algorithm, and support vector machines) to optimize the hyperparameters of a simulation model and obtain the best result efficiently (Choi et al. 2001; Li and Yeh 2002; Yang et al. 2008; Biau 2012; Li et al. 2013; He et al. 2018). However, traditional machine learning algorithms have the ability to optimize simulation models, but they still have some limitations. One of the limitations is that it is hard to get a better simulation accuracy (Lin and Li 2015; He et al. 2018). Another limitation is that when most traditional machine learning algorithms explore each land unit (cell here) in evaluating development suitability according to a set of driving factors, they usually ignore the neighborhood effect with those cells (Li and Yeh 2000; He et al. 2018, Zhai et al. 2020). Some pioneers found that convolutional neural networks (CNNs), which is a deep learning algorithm, can address these limitations (He et al. 2018).

CNNs are a type of artificial neural networks, and the basic network structure of CNN is similar to feedforward artificial neural networks (see Sect. 2.1). CNNs are made up of an input layer, hidden layer(s), and an output layer. However, the hidden layers have three types to build CNN architectures: convolutional layer, pooling layer, and fully connected layer. The basic architecture of CNN is shown in Fig. 6.1. The input of CNN is converted to a tensor that abstracts multiple feature maps with shape information through the convolution layer (LeCun et al. 2015). Pooling layers usually follow convolution layers. The role of pooling layers is to decrease the computational demand and reduce the dimensions of the data. There are two types of pooling: local pooling and global pooling. Local pooling usually combines small clusters, the most

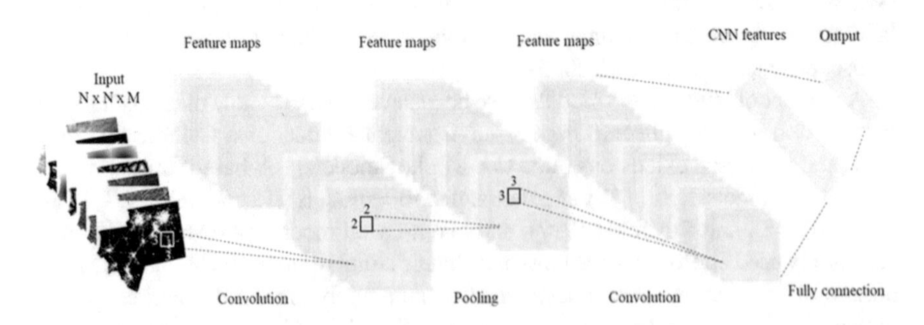

Fig. 6.1 Basic structure of a CNN

common form is the size of 2 × 2. While global pooling with pool size equals to the size of the input. In addition, pooling can be computed using max or average function. Max pooling returns the maximum value from each of the clusters. On the other hand, average pooling returns the average value from the cluster. For example, we have 4 numbers in a 2 × 2 cluster (e.g., 1, 3, 3, 9). It returns 9 when we use max pooling, or 4 if we use average pooling. Each node in the fully connected layer connects every node in the previous layer. The purpose of the fully connected layer is to take advantage of the features from convolution and pooling layers to classify the input into different classes according to the training dataset (Krizhevsky et al. 2012, LeCun, Bengio, and Hinton 2015, Schmidhuber 2015).

The architecture of CNNs is designed to capture the spatial and temporal dependencies from the input through a set of relevant filters (LeCun et al. 2015). In other words, CNNs consider the neighborhood effect for each cell. CNNs have been used in image recognition, computer vision, and natural language processing (LeCun et al. 2015), and a number of scholars from geography-related fields also applied CNNs into geographic studies. For instance, Jean et al. (2016) adopted CNNs to extract information from high-resolution satellite imagery in order to estimate consumption expenditure and asset wealth in Africa. Du et al. (2018) developed a scheme for analyzing measles outbreak using Twitter data and CNN models. Yao et al. (2018) proposed Convolutional Neural Network for United Mining (UMCNN) for mapping urban housing prices using spatial data from different resources at a very fine resolution. At the same time, He et al. (2018) involved CNNs to get better simulation results for urban growth simulation process.

Although the effectiveness and utilities of CNNs in land change models or other geography-related studies have been demonstrated by the pioneers, how to automatically adjust the hyperparameters of CNNs still an open question in those studies. Most of these pioneers mentioned that their future work will investigate how to adjust hyperparameters of CNNs in order to improve the overall accuracy and efficiency. Thus, the purpose of this case study is to examine the capabilities of spatially explicit hyperparameter optimization approach for improving the overall accuracy and efficiency of current deep learning-based (CNNs) spatial simulation models.

Section 6.2 provides a literature review about hyperparameters in CNNs. Section 6.3 presents the study area and data. Section 6.4 discusses the experimental design of this study, which includes the setting of CNNs, CNN-CA model, and implementation. Sections 6.5 and 6.6 present experimental results and discussions. The conclusion is shown in Sect. 6.7.

6.2 Hyperparameters of Convolutional Neural Networks

CNNs are a very powerful machine learning technique, but, overfitting is a serious problem (Srivastava et al. 2014). CNNs involve the idea of ensemble technique, which uses the average result from all possible model configurations as the final result. However, even though ensemble techniques have a strong ability to reduce overfitting,

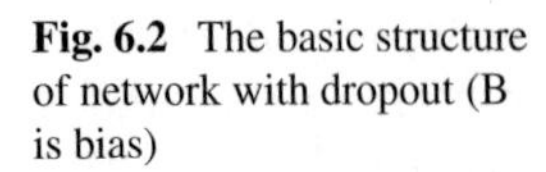

Fig. 6.2 The basic structure of network with dropout (B is bias)

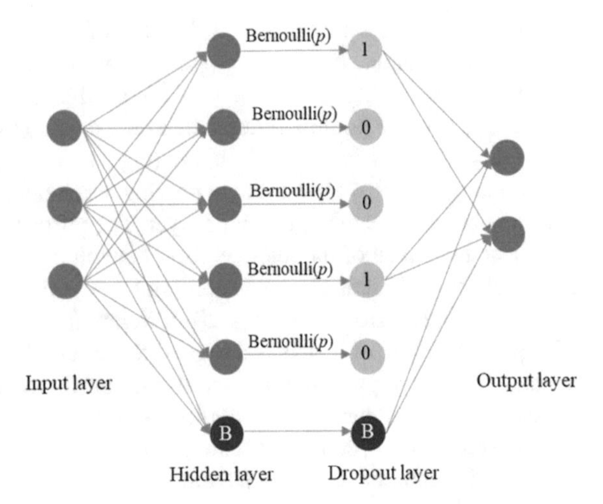

it requires an additional computational expense of training and the high demand for storing multiple models. It is still a challenge for ensemble techniques if the models are large and complicated (Salakhutdinov and Mnih 2008; Xiong et al. 2011; Srivastava et al. 2014). Dropout is an important hyperparameter in deep learning that can address overfitting issues, expensive computation, and large storage space. Dropout was first introduced by Hinton and Krizhevsky (Hinton et al. 2012; Krizhevsky et al. 2012), and then Srivastave et al. (2014) further discussed how to use dropout to reduce overfitting. The mechanism of dropout is that it randomly and temporarily removes one or more units from the network in the training process. In the testing process, we use all units in the networks and trained weights to get the results based on the testing dataset. The trained weights are scaled-down by the "removing probabilities". For example, if we run the network 10 times and a unit is removed 5 times, the outgoing weight of this unit will be multiplied 0.5 ((10–5)/10).

The dropout refers to the percentage of dropping out units. I use dropout of 0.6 as an example to show the working process of the dropout in neural network models (Fig. 6.2). In this example, the number of hidden nodes is 5 and the dropout is 0.6, which means 3 (5 * 0.6) hidden nodes will be temporarily removed from the network. The independent Bernoulli random variables[1] assign to each unit and each of them has probability p of being 1 (i.e., how many units will be used in the network). Besides the overfitting issue, dropout can handle the computational challenge for ensemble techniques because the architectures of networks are different for each iteration (Hinton et al. 2012; Srivastava et al. 2014). Please refer to Srivastava (2014) for more information about dropout.

Although dropout significantly reduces the overfitting issue in CNNs, starting with a high learning rate, then it decays during the learning process provided a significant boost in model performance (Hinton et al. 2012). Moreover, dropout

[1] A Bernoulli random variable is the simplest kind of random variable that has only two possible outcomes. It takes on 1 if an experiment is success with probability p and 0 otherwise.

with regularization methods (e.g., L2, max-norm, and KL-sparsity) gives a lower generalization error (Srivastava et al. 2014). Thus, in this study, I also involve a regularization method (L2 here) to further reduce overfitting issue. L2 regularization, commonly called weight decay, is one of the most used regularization methods. The term "weight decay" refers to for each weight update, the weights are multiplied by a small value, which prevents the weights from growing too large. The updated weight formula (Zheng et al. 2019) with weight decay is listed as follows:

$$\Delta w_{ij}^{l}(n) = \lambda[-\delta \frac{\partial E(n)}{\partial w_{ij}^{l}(n)} + \alpha \Delta w_{ij}^{l}(n-1)] \tag{6.1}$$

where $-\delta \frac{\partial E(n)}{\partial w_{ij}^{l}(n)} + \alpha \Delta w_{ij}^{l}(n-1)$ is the original updated weight formula, see Zheng et al.'s study for detailed explanation (Zheng et al. 2019). λ is the weight decay coefficient.

The reasonable values of weight decay coefficient are between the range of 0 and 0.1 (Kuhn and Johnson 2013). However, some studies suggested that the values of weight decay coefficient should be small enough. For example, Reed and Marksll chose the value of 0.001 because it is the most used number (Reed and MarksII 1999). Krizhevsky et al. found that weight decay of 5×10^{-4} was important for the model to learn using ImageNet dataset (Krizhevsky et al. 2012). Chollet (2017) also used ImageNet dataset, but he found that the value of 4×10^{-5} was better than the value of 1×10^{-5}. Therefore, it is important to select a suitable weight decay coefficient specific to the network and datasets. Because if the value of weight decay coefficient is very large, it will lead to underfitting.

6.3 Study Area and Data

Lower High Rock Lake Watershed area (HRLW) is the study area. The lower HRLW covers eight counties in North Carolina, including Davie, Forsyth, Davidson, Rowan, Cabarrus, Randolph, Guilford, and Iredell. The total area of this region is 10,823 km^2. Forest is the major land cover type in the lower HRLW, percentage of forest is 49.07, 47.71, 46.78, and 47% in year 2001, 2006, 2011, and 2016. Although forest covers almost half of the area in the lower HRLW, Forsyth and Guilford are the most urbanized counties in this region (See Fig. 6.3 and Table 6.1).

The datasets used in this study were collected from U.S. National Land Cover Database (NLCD; https://www.mrlc.gov/data). The spatial resolution of these four land cover is 1-arc second or around 30 m. In order to simplify the model, I further reclassified those 16 classes into five land cover types: farmland, natural, forest, urban, and water. The datasets of major roads, city centres, and county boundaries were retrieved from U.S. Census Bureau (https://www.census.gov/data.html). The dataset of streams (including stream network and watershed data) was obtained from USDA National Hydrography Dataset (https://www.usgs.gov/core-science-systems/

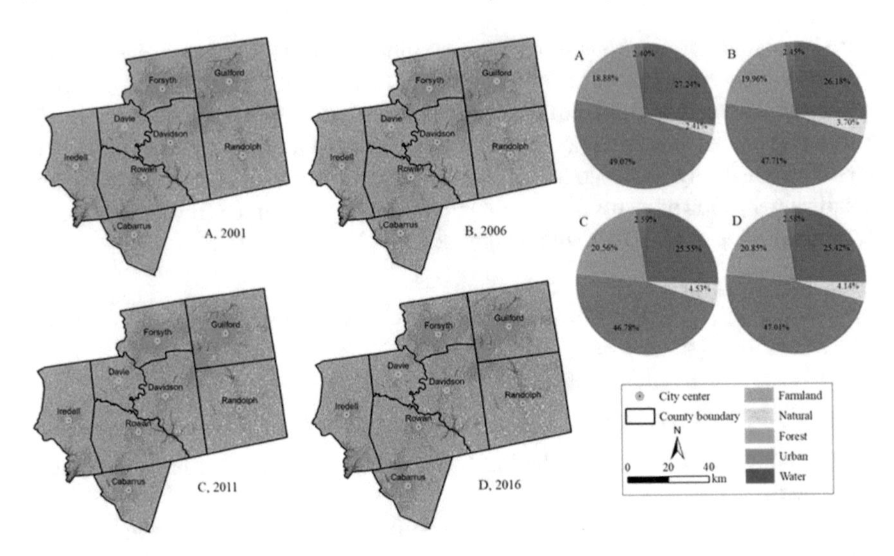

Fig. 6.3 Map of the study area

ngp/national-hydrography/access-national-hydrography-products). Elevation data was collected from NASA Shuttle Radar Topography Mission (https://www2.jpl.nasa.gov/srtm/). The landscape size is 4,378 and 4,746 in terms of number rows and columns.

Using the statistics for urban areas in lower HRLW, each county's urban proportion and expansion rate are calculated and shown in Table 6.2. As the table shows, Cabarrus County had the largest growth rate from 2001 to 2016 (4.95%), followed by Guilford, Forsyth, and Iredell counties, which increased by 3.38%, 2.88%, 2.08% respectively. Cabarrus County had the largest expansion rate from 2001 to 2006 and 2011–2016, while Guilford had the largest expansion rate from 2006 to 2011. The average urban expansion rates of lower HRLW for these four periods increased from 19.06 to 21.15%. However, the urban expansion rate of Randolph County was the lowest from 2001 to 2006 and 2011 to 2016, while Davie County had the lowest urban expansion rate from 2006 to 2011. Meanwhile, due to the impact of the 2006–2011 economic recession, the urban expansion rates for each county were lower than the previous period.

In this study, I use seven driving factors to decide whether the land converts into urban area, which refers Tang and Yang's study (2020), including elevation, slope, stream density, distance to streams, distance to major roads, distance to city center, and development pressure (see Fig. 6.4). The development pressure calculates the percentage of urban areas within a neighborhood (neighborhood size is 1050 m by 1050 m). The stream density derived from a search radius of 500 m. I use Euclidean distance to get the proximity driving factors (i.e., distance to major roads, distance to streams, and distance to city center).

Table 6.1 Land cover and land change data for each county (unit: km^2)

Land cover	Year	Cabarrus	Davie	Davidson	Forsyth	Guilford	Iredell	Randolph	Rowan	Total
Farmland	2001	246.12	251.66	358.34	177.21	405.40	548.83	523.15	437.27	2947.99
	2006	231.27	245.41	342.75	162.99	383.15	532.46	507.24	427.41	2832.68
	2011	224.24	242.08	333.96	155.17	369.17	522.70	495.88	421.42	2764.60
	2016	220.40	241.33	332.86	153.79	366.51	521.62	494.22	420.61	2751.33
Natural	2001	25.50	12.04	44.15	15.02	18.84	29.37	87.83	27.70	260.46
	2006	44.10	16.03	59.98	23.30	42.71	45.58	128.46	39.92	400.08
	2011	46.49	24.18	66.75	30.40	59.10	52.79	160.02	50.00	489.73
	2016	33.63	23.90	56.94	33.87	56.52	54.73	139.58	48.52	447.70
Forest	2001	452.10	354.72	800.58	468.42	701.28	681.98	1195.87	654.35	5309.29
	2006	420.40	353.97	788.33	457.37	666.88	666.32	1164.71	644.46	5162.45
	2011	417.32	348.46	782.66	447.66	643.73	656.59	1130.90	634.63	5061.94
	2016	422.61	347.63	791.44	442.19	644.18	651.89	1151.45	635.39	5086.80
Urban	2001	200.27	59.99	221.60	390.17	531.17	228.69	214.87	196.59	2043.35
	2006	227.79	62.71	231.83	407.07	563.67	244.49	220.17	202.70	2160.43
	2011	235.46	63.36	236.76	417.84	583.57	257.30	224.77	206.10	2225.17
	2016	246.93	65.17	239.27	420.98	588.68	260.79	226.58	208.22	2256.61
Water	2001	19.18	13.35	44.07	17.95	46.05	56.48	22.18	40.60	259.88
	2006	19.60	13.64	45.84	18.06	46.34	56.51	23.31	42.02	265.31
	2011	19.65	13.70	48.61	17.71	47.18	55.98	32.33	44.36	279.52
	2016	19.60	13.73	48.22	17.96	46.85	56.32	32.07	43.77	278.52

Table 6.2 Urban proportion for each county (Note: (2006–2001)%, expansion ratio from 2001 to 2006; (2011–2006)%, expansion ratio from 2006 to 2011; (2011–2001)%, expansion ratio from 2001 to 2011)

	2001 (%)	2006 (%)	2011 (%)	2016 (%)	(2006–2001) (%)	(2011–2006) (%)	(2016–2011) (%)	(2016–2001) (%)
Cabarrus	21.23	24.15	24.97	26.18	2.92	0.81	1.22	4.95
Davie	8.67	9.07	9.16	9.42	0.39	0.09	0.26	0.75
Davidson	15.09	15.78	16.12	16.29	0.70	0.34	0.17	1.20
Forsyth	36.51	38.09	39.10	39.39	1.58	1.01	0.29	2.88
Guilford	31.20	33.10	34.27	34.57	1.91	1.17	0.3	3.38
Iredell	14.80	15.82	16.65	16.88	1.02	0.83	0.23	2.08
Randolph	10.51	10.77	11.00	11.09	0.26	0.23	0.09	0.57
Rowan	14.49	14.94	15.19	15.35	0.45	0.25	0.16	0.86
Average	19.06	20.22	20.81	21.15	1.15	0.59	0.34	2.08

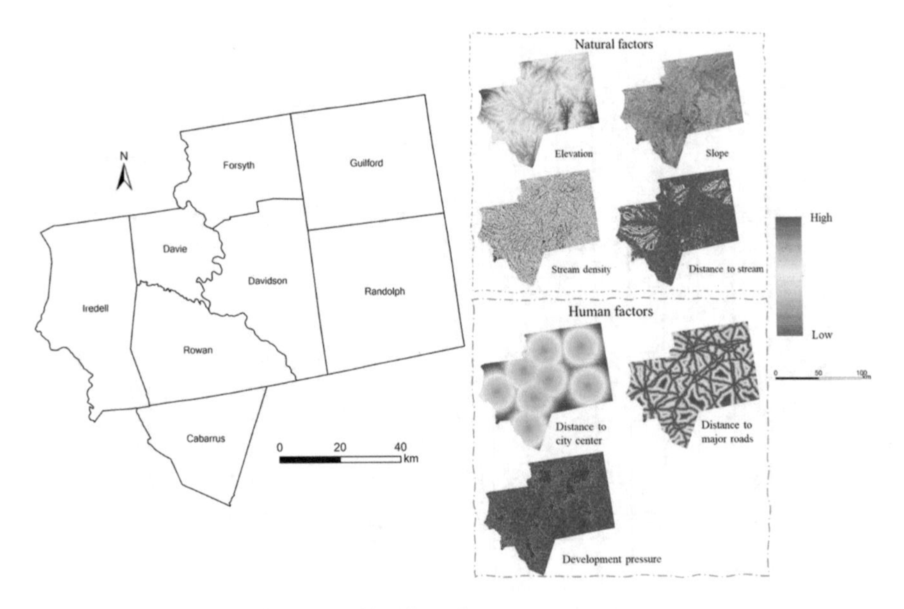

Fig. 6.4 Selected driving factors used in this study

6.4 Experimental Design

The urban growth data were obtained from historical data, and I focus on examining the transformation of non-urban areas to urban areas. Thus, I aggregated five land use types into two categories: 0 and 1. Code 0 means unchanged, including non-urban areas (farmland, natural, forest and water) converted to non-urban areas, and urban areas converted to non-urban areas; code 1 means non-urban areas converted

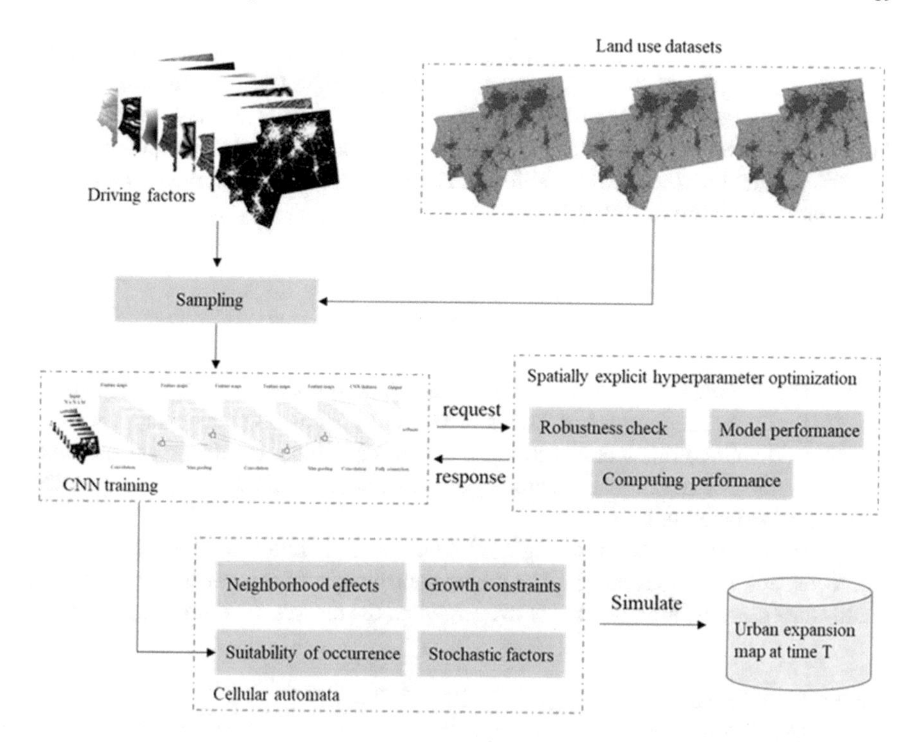

Fig. 6.5 Flowchart of urban land change simulation through CNN, CA and spatially explicit hyperparameter optimization. T stands for the simulated year

to urban areas. I excluded the unchanged urban areas in the previous period (e.g., for simulating the urban land change from 2001 to 2006, we removed urban cells in 2001). In this study, I adopt the CNN framework to calculate the urban development probability. Figure 6.5 describes the flowchart of the CNN-CA model. This study is divided into the following steps. (1) I generate driving factors with the consideration of neighborhood information. (2) The CNN model is trained, and then I use spatially explicit hyperparameter optimization approach to find the appropriate hyperparameters in order to get the best model performance. (3) The overall development probability is derived from the CNN-CA model and then simulate future urban patterns.

6.4.1 Setting of CNN Model

The CNN model used in this study has nine layers, including one input layer, seven hidden layers, and one output layer. More specifically, there are one input layer with seven nodes (links with the input variables), three convolution layers, two max pooling layers, one fully connected layer, one softmax layer, and one output layer with two nodes (links with the two outcomes). Table 6.3 presents detailed information of the hidden layers. Figure 6.6 shows the structure of the CNN model used in this study. I use rectified linear units (ReLU) activation function in these convolution layers. Compared with sigmoid (logistic) activation function and hyperbolic (tanh) activation function, ReLU has the ability to avoid saturation (Goodfellow et al. 2016). The general equation of ReLU is:

$$f(x) = \begin{cases} x, & x \geq 0 \\ 0, & x < 0 \end{cases} \tag{6.2}$$

where $f()$ is the function of ReLU, and x is the random variable. From this equation, ReLU is a linear function when values of random variable is greater than zero. As mentioned by Goodfellew et al. (2016), ReLU has a number of advantages of linear models, such as a good generalization performance and a simple way to optimize gradient-based methods.

Typically, softmax layer or sigmoid function is adopted in the last layer of CNN (except the output layer). The major contribution of this layer is to calculate multi-class classification results. However, when the classes are mutually exclusive,

Table 6.3 The information of hidden layers used in this study

Layer	Feature map	Note
1st convolution layer	48 × 48 × 16	3 × 3 convolution kernel
1st max pooling layer	24 × 24 × 16	Size of 2 × 2
2nd convolution layer	22 × 22 × 32	3 × 3 convolution kernel
2nd max pooling layer	11 × 11 × 32	Size of 2 × 2
3rd convolution layer	9 × 9 × 32	3 × 3 convolution kernel
Fully connected layer		96 nodes
Softmax layer		Probabilities of outcomes sum to 1

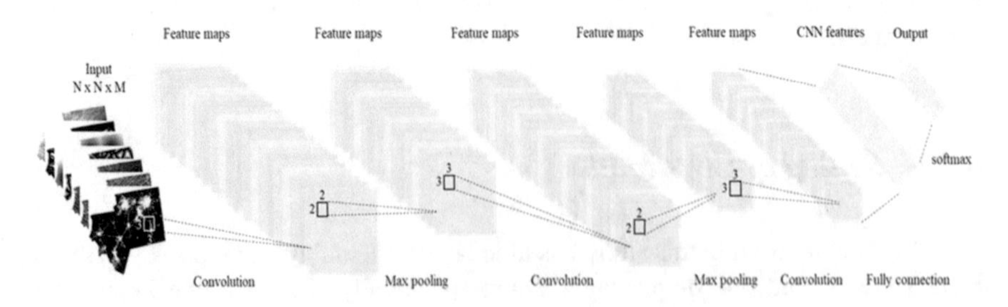

Fig. 6.6 Structure of CNN model used in this study

softmax layer is better than sigmoid function. One of the reasons is that the probabilities of outcomes sum to 1 for softmax layer. Another reason is that sigmoid function exists gradients vanish issue when the probability gets close to 0 or 1 (LeCun et al. 2015; Goodfellow et al. 2016). In this case study, the results are mutually exclusive, i.e., non-urban to urban or non-urban to non-urban, thus, softmax layer is adopted in here. Moreover, I involve weight decay and dropout in the CNN model in order to avoid overfitting issue.

As suggested by Srivastava et al. (2014), dropout should use 10–100 times the learning rate, and the better values of momentum should be around 0.95–0.99. Based on their suggestions and my preliminary experiment, I adopt the value of 0.001 as the learning rate and value of 0.9 for momentum. The values of dropout and weight decay range from 0 to 1 and from 0 to 0.001, respectively. In the training process of the CNN model, batch size is also an important hyperparameter of the model training. The best training stability and generalization performance usually along with smaller batch size, whereas larger batch sizes may accelerate the convergence. Based on previous studies' suggestions (Bengio 2012; Zhai et al. 2020), I adopt the batch size of 32 in this study.

6.4.2 CNN-Based Cellular Automata

Neighborhood effect (*NE*) is one of the essential components in exploring spatial process (First Law of Geography (Tobler 1970)) and a part of the traditional CA model. Besides neighborhood effect, the remaining three parts are the suitability of occurrence (*OP*), growth constraints (β), and stochastic factor (α) (Wolfram 1983; Li and Yeh 2002).

The overall development suitability is usually based on related biophysical and socioeconomic factors through logistic regression. However, with the advances in computing techniques, a set of machine learning models were adopted to determine the coefficients of those driving factors because the spatial process is complex and nonlinear, such as artificial neural networks and random forest. The neighborhood effect is a key component in CA model. Von Neumann neighborhood and Moore neighborhood are the commonly used ways to examine the neighborhood effect (Batty et al. 1997). Von Neumann neighborhood is a diamond shape neighborhood (2-dimensional space), whereas Moore neighborhood is a square shape neighborhood (2-dimensional space). Figure 6.7 describes Von Neumann neighborhood and Moore neighborhood with first-order zone. The neighborhood effect is defined as:

$$NE_{t,i} = \frac{total\ number\ of\ urban\ cells\ within\ the\ neighborhood}{n \times n - 1} \tag{6.3}$$

where the neighborhood region is depicted by a n×n square regions. $NE_{t,i}$ is the neighborhood effect of cell i at time t. The focal cell i does not belong to the part

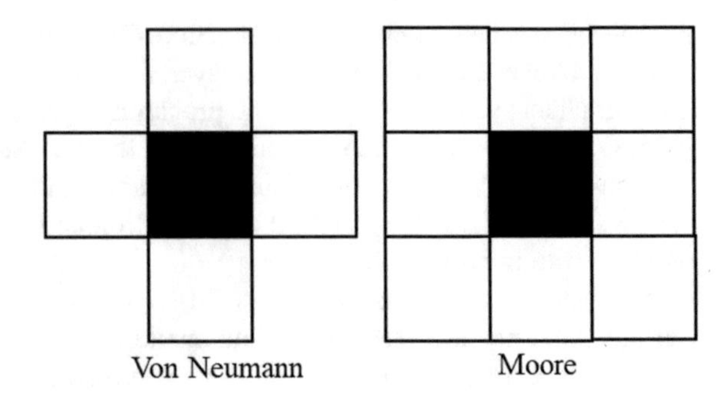

Fig. 6.7 Von Neumann neighborhood and Moore neighborhood with the first-order zone (square with black color is focal cell; square with white color is neighbor of the focal cell)

of estimating the neighborhood effect on urban growth, thus, we should remove this cell in estimating neighborhood effect.

Growth constraint is used to control the unallowable transition, for example, water areas are usually not allowed to convert to other types of land use. In this study, the growth constraint is a binary variable. 0 denotes the cell is water, and 1 denotes the cell is other types of land use. The purpose of stochastic factor is to maintain the randomness of the transition process. The range of stochastic factor in this study is a random number in [0, 1] follow the uniform distribution. Overall, the transition probability is defined as:

$$P_{t,i} = PO_{t,i} \cdot NE_{t,i} \cdot \beta_{t,i} \tag{6.4}$$

where $P_{t,i}$ is the transition probability of non-urban cell i at time t convert to urban cell at time $t + 1$. $PO_{t,i}$ is the suitability of occurrence of cell i at time t, and $NE_{t,i}$ is the neighborhood effect of cell i at time t. $\beta_{t,i}$ is a growth constraint of cell i at time t.

Specifically, the CA part of the CNN-CA model is shown in Fig. 6.8. First, I estimate the suitability of occurrence through the CNN model. And then I find the values of neighborhood effect based on 5×5 Moore neighborhood and growth constraints for cell i. Next, the transition probability can be estimated using Eq. 6.4. Then, the transition probability of cell i is compared with a random number to determine whether the cell converts to a urban cell. If the transition probability of cell i is greater than the random variable, cell i will convert to urban cell at time $T + 1$.

6.4.3 Implementation

The script of this CNN-CA model was written in Python programming language using Pytorch library (https://pytorch.org/), the scripts of preprocessing and post-processing were also written in Python programming language. ESRI ArcGIS Pro

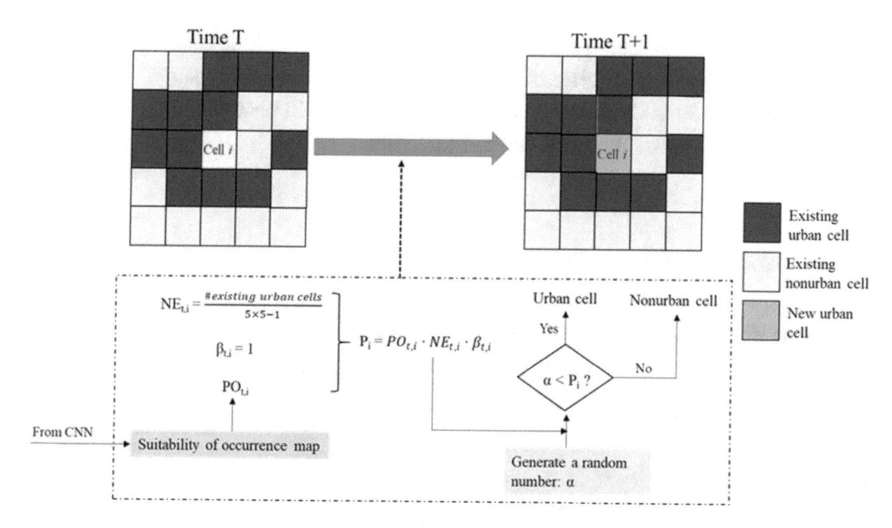

Fig. 6.8 The processing architecture of CA ($P_{t,i}$ is the transition probability of non-urban cell i at time T convert to urban cell at time $T + 1$. $PO_{t,i}$ is the suitability of occurrence of cell i at time T, and $NE_{t,i}$ is the neighborhood effect of cell i at time T. $\beta_{t,i}$ is the growth constraint of cell i at time T)

processed all GIS data. Euclidean distance tool from ArcGIS was used to calculate all distance-based variables. Our CNN model with spatially explicit hyperparameter optimization was deployed on a Redhad Linux-based high performance computing environment at University Research Computing (URC; https://urc.uncc.edu). Specifically, Copperhead cluster was used for this study.

6.5 Results

6.5.1 Accuracy Assessment

To evaluate the simulation results, the similarity metrics based on confusion matrix were used. Most previous studies adopted overall accuracy, Kappa coefficients, producer accuracy, user's accuracy at the cell level (Fielding and Bell 1997; Li et al. 2008). In this study, I adopted Figure of Merit (FoM). FoM is "a ratio of the intersection of the observed change and predicted change to the union of the observed change and predicted change" (Page 22, (Pontius et al. 2008)). The formula of FoM is shown in the following equation (Pontius et al. 2008):

$$\text{FoM} = \frac{B}{A + B + C + D} \tag{6.5}$$

where, for all observed change cells, A is the total number of the cells that simulate unchange, B is the total number of the cells that simulate change, and C is the total number of the cells that simulate to wrong category. For all observed unchange cells, D is the total number of the cells that simulate change.

6.5.2 Model Performance

To evaluate the model performance of spatially explicit hyperparameter optimization approach, I use cross-entropy (Eq. 6.6) loss function to optimize the CNN model and accuracy (Eq. 6.7) to measure the performance of the model. The equations are shown as follow:

$$\text{cross\_entropy} = -\sum_{i=1}^{n}\sum_{j=1}^{m} y_{i,j}\log(P_i) \tag{6.6}$$

$$\text{accuracy} = \frac{A_c}{A_c + A_u} \tag{6.7}$$

where $y_{i,j}$ stands the observed value that cell i belongs to class j, and P_i denotes the probability of predicted sample i belonging to class j. A_c stands for the number of correctly simulated cells, A_u is the number of incorrectly simulated cells.

Figure 6.9 illustrates the learning curve and scatterplot of values of accuracy (generalization performance) for the conventional EA-based hyperparameter optimization and spatially explicit hyperparameter optimization. These scatterplots showed the distributions of original sampled points, the middle generation of the evolutional process, and the last generation that has the convergent results (i.e., the error difference among five generations is less than 1%). The scatterplots of values of accuracy visualize the training process. For those two approaches, the lowest accuracy concentrates around the upper right and right regions during the entire process. The higher value of accuracy is caused by relatively smaller dropout (under 0.5) and relatively smaller weight decay (under 6×10^{-4}). In general, the generalization performance of CNN model is high (i.e., high value of accuracy) when both dropout and weight decay are low. In the meanwhile, spatially explicit hyperparameter optimization uses less computing time to get the results.

6.5.3 Generalization Performance of Hyperparameters

Given the values of the accuracy of sampled hyperparameters, a continuous surface of the values of accuracy was conducted using spatial interpolation methods. Figure 6.11

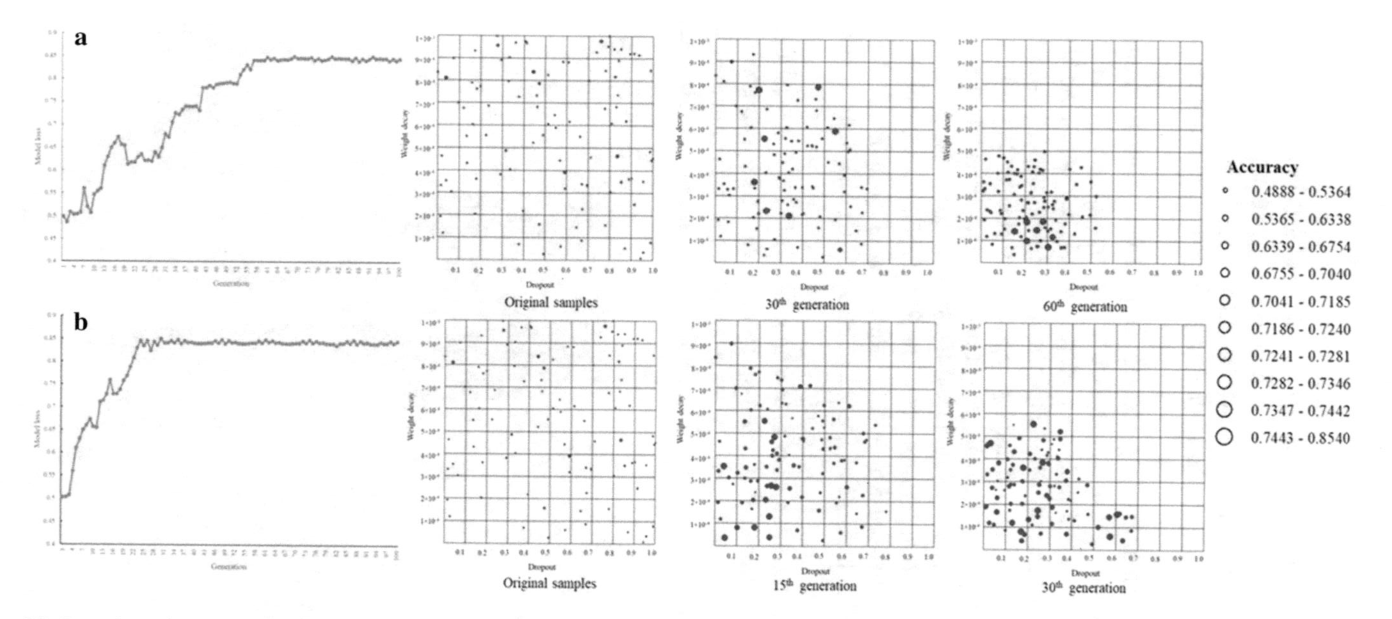

Fig. 6.9 Learning curve and scatterplots of values of accuracy for hyperparameter sets over different generations (A: red circle is sampled points that generated by conventional EA-based hyperparameter optimization; B: blue circle is sampled points that generated by spatially explicit hyperparameter optimization; circle size is proportional to MSE, circle size from small to large stands for low MSE to large MSE; outliers were excluded)

shows the generalization performance maps of the values of accuracy using conventional EA-based approach and spatially explicit approach. In the processes of both EA-based approaches, the highest accuracy concentrates in the lower-left corner region, particularly when dropout is less than 0.6 and weight decay is less than 4×10^{-4}. However, when dropout ranges from 0.8 to 1, the accuracy of the model achieves the lowest values for both approaches. In general, the generalization pattern of hyperparameter space is similar to the conventional approach and the approach proposed in this book.

From Fig. 6.11, the high accuracy occurs as dropout and weight decay decrease. A relatively larger dropout (above 0.8) leads to the lowest accuracy. The small weight decay (below 4×10^{-4}) and small dropout (below 0.5) can obtain the best model performance (highest accuracy here). The patterns of conventional EA-based hyperparameter optimization approach and my approach are similar (Fig. 6.11a, b). That is, dropout ranges from 0 to 0.5 and weight decay ranges from 0 to 4×10^{-4} lead to high generalization performance. I used RMSE to measure the prediction errors of these two generalization performances. The RMSE of cross-validation is 7.56×10^{-3} for conventional EA-based hyperparameter optimization approach, 3.15×10^{-3} is for spatially explicit hyperparameter optimization. However, some regions show different patterns.

Figure 6.10 illustrates the map of a continuous pattern of standard errors in response to hyperparameter sets based on spatially explicit hyperparameter optimization. The value of standard error was calculated based on the values of accuracy for each sampled hyperparameter set. Standard error serves as a measure of variation,

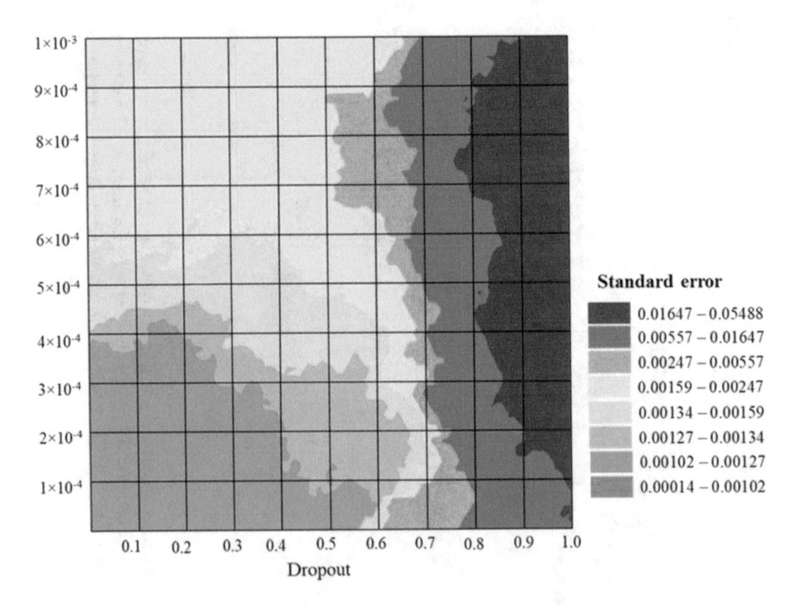

Fig. 6.10 Distribution of standard error based on the values of accuracy using spatially explicit hyperparameter optimization methods

the smaller the standard error, the result is more accurate and the sample population is more representative of the overall population. The RMSE for cross-validation for the standard error is 9.17×10^{-3}. The results of standard errors show that my approach is robust and accurate (most values of standard error are less than 1.6×10^{-2}). The generalization performance is becoming worst when dropout and weight decay are large. Specifically, the larger standard errors occur when dropout between 0.7–1. When dropout is less than 0.6 and weight decay is less than 4×10^{-4}, the results of hyperparameter optimization are the most stable. However, relatively larger standard errors occur when weight decay ranges from 4×10^{-4} to 1×10^{-3}. In general, smaller dropout and weigh decay have stable generalization performance.

6.5.4 Prediction Performance

From the results of hyperparameter optimization, I adopt the hyperparameter set with the highest accuracy (dropout: 0.18; weight decay: 1×10^{-4}) to train the CNN model. The CNN model is established through data from 2001–2006, and the data from 2006–2011 is used to validate the model. In the training process, I adopted a stratified random sampling method to selected 40,000 sampled cells (20,000 for urban, 20,000 for nonurban) to fit the model. Then, 80% of the sample data is used as the training dataset and remaining 20% sample data is employed as the testing dataset. The accuracy and loss values are calculated using the testing dataset. During the urban expansion process, the land use type of the cell's neighborhood determines its transition state, that is, within a neighborhood, the closer to the cell, the greater effect on the state transition.

Due to the size of the dataset (around 600G), I ran the simulation model in a single county. Forsyth County was selected for validation because this county was one of the most urbanized counties in the lower HRLW, and the urban growth ratio also was top-ranking in 2001–2006, 2006–2011, and 2011–2016 periods. The landscape size of Forsyth is 1,438 × 1,223 in terms of number rows and columns, and I removed the cells that are urban type in the previous period. For example, if a cell in 2006 is an urban type, I assume this cell will not convert to other land use types (urban or non-urban) in 2011. Moreover, another model (logistic regression-CA) was compared to test the utilities of the CNN-CA model. In order to ensure the data consistency, the logistic regression adopt the same sample data as CNN model, that is, I use the same training and testing datasets to construct those two models. I adopt 0.4 as a threshold of logistic regression (i.e., simulated probability is greater than 0.4, the non-urban cell will transform to urban cell). The overall development probability usually derives based on related biophysical and socioeconomic factors through CNN. The map of development probability is shown in Fig. 6.12.

The results of validation accuracy are shown in Table 6.4. As shown in Table 6.4, the overall accuracy of logistic regression achieves to 75.39%, and the overall accuracy of CNN is 90.79%. Meanwhile, the accuracy of identifying non-urban to urban cells for CNN is higher than logistic regression. CNN model can identify 78%

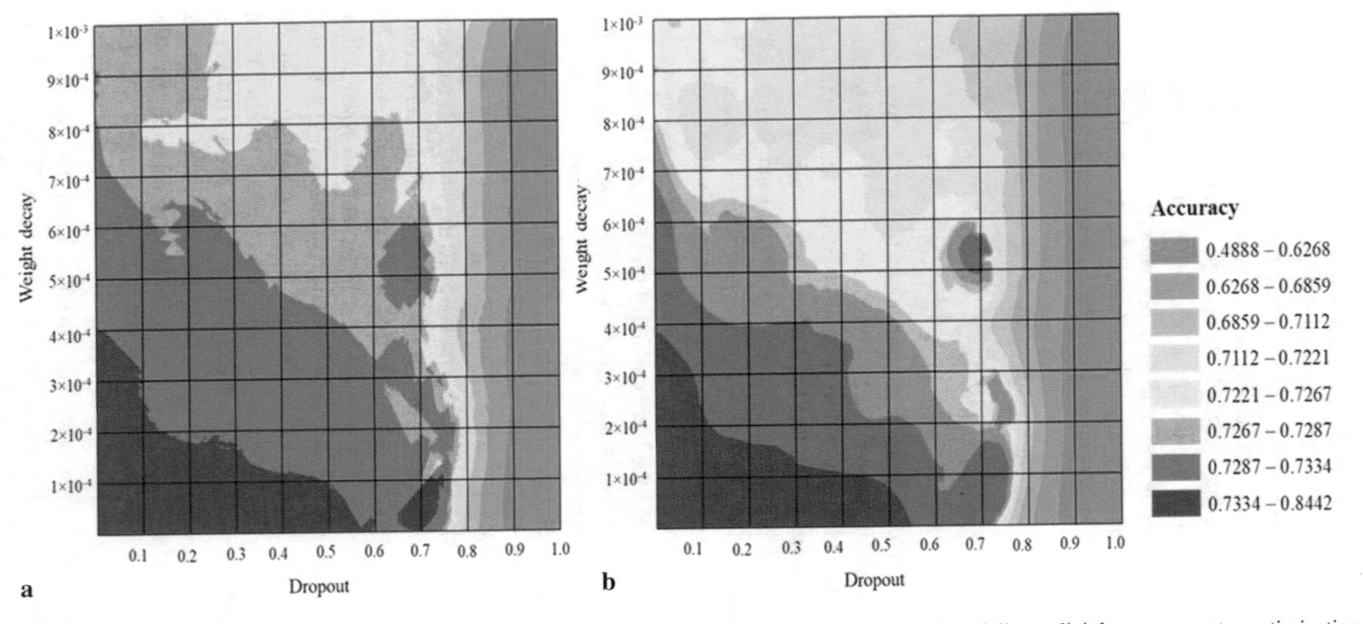

Fig. 6.11 Maps of generalization performance for conventional EA-based hyperparameter optimization and spatially explicit hyperparameter optimization (a: conventional EA-based hyperparameter optimization using the 60th generation result; b: spatially explicit hyperparameter optimization)

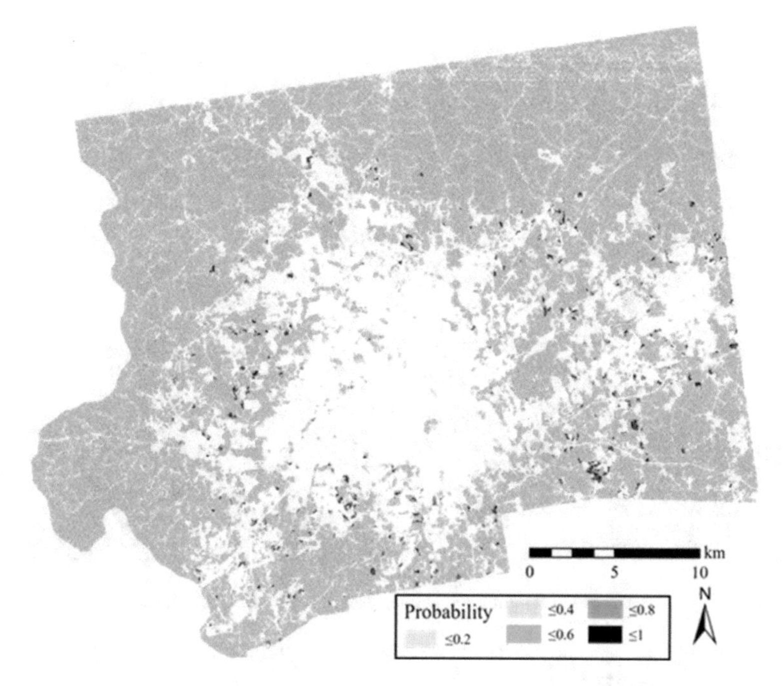

Fig. 6.12 Probability-of-occurrence of urban growth changes for 2006–2011 period

Table 6.4 The accuracy of the validation results via logistic regression and CNN (0, non-urban cells convert to non-urban cells; 1, non-urban cells convert to urban cells)

Results	Logistic regression (%)	CNN (%)
Accuracy of 0 s as 0 s	76.7	91
Accuracy of 1 s as 1 s	2.4	78
Overall accuracy	75.39	90.79
FoM	0.04	1.23

of the non-urban to urban cells, whereas logistics regression can identify 2.4% of non-urban to urban cells. Furthermore, logistic regression only correctly identifies 76.7% of non-urban to non-urban cells, but the accuracy of identifying non-urban to non-urban cells for CNN is over 90%.

Figure 6.13 shows the maps of observed and simulated land change patterns for 2006–2011 period, and the areas with white color mean the cells were urban in 2006. From these maps, we can see that most non-urban cells remain persistent. Although the urban growth rate in Forsyth County was top-ranking in the study area, land change quantity (exclude urban cells in the previous period) only occupies about 1.63% and 0.48% of the entire county in 2006–2011 and 2011–2016 periods. The validation accuracy for both land change types (non-urban to urban and non-urban to non-urban) are high even though most regions do not experience land change. Furthermore, I simulate the urban growth for 2011–2016 period. Figure 6.14 shows

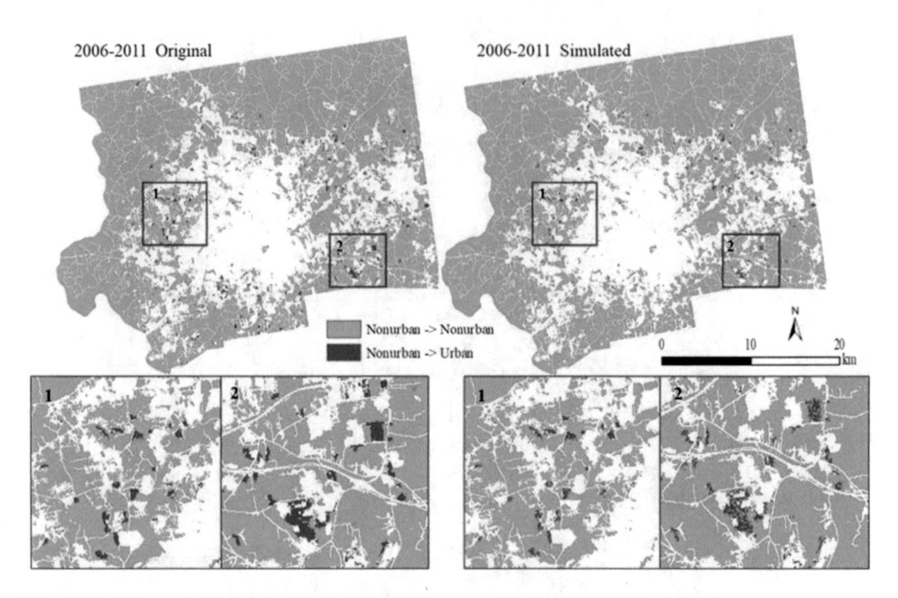

Fig. 6.13 Maps of observed and simulated urban growth for 2006–2011 period

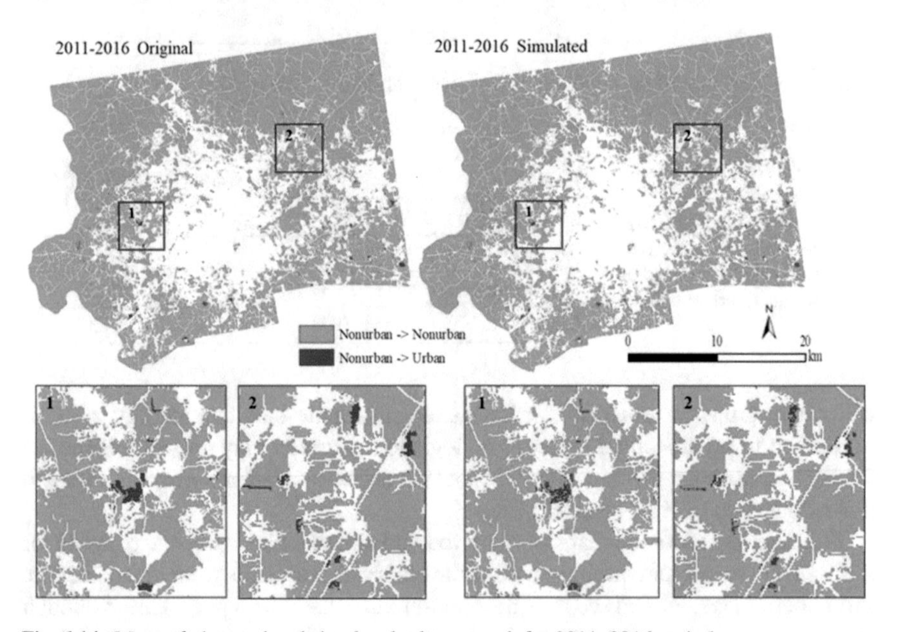

Fig. 6.14 Maps of observed and simulated urban growth for 2011–2016 period

the simulated urban growth for 2011–2016. The total changed cells for urban areas was 3,480, and my simulated result got 2,436 non-urban cells to urban cells. The simulation accuracy of non-urban to urban is 0.70, and the simulation accuracy of non-urban to non-urban is 0.997. The overall simulation accuracy is 0.9956 and FoM is 0.34%.

6.5.5 *Parallel Computing Performance*

The spatially explicit hyperparameter optimization approach randomly generated 100,000 sampled hyperparameter sets (100 sampled hyperparameter sets for each generation). For each generation, each CPU handled only one sampled hyperparameters. Thus, I adopted 100 CPUs in this study. The estimated sequential computing time of spatially explicit hyperparameter optimization is around 5,169,630 s (about 59.83 days). However, the parallel computing time of conventional EA-based hyperparameter optimization approach was around 692,381 s (about 8 days). The speedup was 7.47. Moreover, the hyperparameter approach used in this study can further reduce the computing time (Fig. 6.9). The estimated parallel computing time was 207,714 s (around 2.4 days), where speedup achieves to 24.89.

6.6 Discussions

6.6.1 *The Simulation Performance of CNN-CA Model*

Simulation models involved deep learning techniques, and these have been a hot research topic in the GIScience community. Some pioneers already discussed the ability of deep learning techniques in spatial simulation models. One benefit of deep learning-based spatial simulation models is that it considers the neighborhood information (i.e., neighborhood effects) of the driving factors. Also, the simulation accuracy increases compared to other models, such as artificial neural networks and random forests. However, how to further improve model performance (i.e., find the optimal hyperparameters atomically) is an open question in current research (He et al. 2018; Zhai et al. 2020). Due to the complexity of deep learning techniques, hyperparameters of deep learning techniques will affect the computing performance and validation accuracy. The purpose of this case study is to demonstrate the utility of spatially explicit hyperparameter optimization approach in deep learning-based simulation models.

As evidenced by comparing two hyperparameter optimization approaches, the best overall generalization performance (validation accuracy) using spatially explicit hyperparameter optimization is around 0.86. In contrast, the best generalization performance using the conventional approach is around 0.84. Furthermore, the standard errors indicate that the results of the spatially explicit approach are robust and accurate (Fig. 6.10). Although these two approaches have very close generalization performances, the spatially explicit approach uses less time. Moreover, the approach proposed in this book provides an answer to the question mentioned above (i.e., how to find the optimal hyperparameters automatically). In general, the spatially explicit hyperparameter optimization approach could provide accurate results with less computing time.

I compare the validation accuracy of CNN model with logistic regression (Table 6.4). My CNN model has a better overall performance than logistic regression (overall accuracy of CNN is 90.79%, and overall accuracy of logistic regression is 75.39%). Moreover, CNN model has a better performance than logistic regression when identifying non-urban to urban cells. Although the study area is not highly urbanized (the percentage of the urban area was 39.39% in 2016) and most of non-urban cells do not convert to urban cells, the CNN model still can correctly identify 78% changed cells (non-urban to urban cells). FoM is used to measure the amount of variation in the urban growth process, and the overall accuracy focuses on the amount of correctly identify cells. The validation accuracy of FoM is 1.23%, which is a small number. However, land change quantity (non-urban to urban; exclude urban cells in the previous period) only occupies about 1.63% of the entire county in 2006–2011 period. Thus, for a low net land change area, a single accuracy measurement cannot fully explain the model performance. Moreover, the map of simulated urban growth (Figs. 6.13 and 6.14) determined by the CNN-CA model is mostly consistent with the actual patterns, particularly for the spatial distribution. Some cells are exactly the same as the actual pattern. Therefore, the results indicate that the CNN-CA model improves the simulation accuracy of a low net land change study area.

6.6.2 Computing Performance

In Chap. 5, I already showed that spatially explicit hyperparameter optimization could expedite convergence and reduce computing time for ANN-based spatial models. However, in this study, I applied my approach to deep learning-based spatial models, the performance of our approach is still significantly effective.

The results showed that this approach can automatically find the optimal hyperparameters for deep learning techniques and effectively speed up the computing time. The spatially explicit hyperparameter optimization considers spatial dependence in the search space, which significantly accelerates the search process (about 70 generations down to about 30 generations; Fig. 6.9). The entire running time decreased to 207,714 s (estimated value; about 57.7 h). This approach significantly handles computational intensity. Furthermore, the results showed that prior knowledge (the results from spatial interpolation) could expedite convergence and reduce the computing time (Xiao et al. 2002).

Meanwhile, this approach answered the open question that exists in current deep learning-based spatial simulation models, that is, how to avoid increasing the overall running time when involving hyperparameter optimization. By adopting my approach, it can not only handle the computational intensity issue in deep learning and hyperparameter optimization, but it also can improve the entire model performance without introducing subjective factors.

6.7 Conclusion

In this study, I proposed a CNN-CA spatial simulation model with integration of spatially explicit hyperparameter optimization approach. The result of hyperparameter optimization demonstrated the spatially explicit hyperparameter optimization can help CNN to find the appropriate settings within a short time. My approach also addressed the challenge of parameter setting in current CNN-based spatial simulation model. The simulation results showed that CNN-CA model has better performance than logistic regression-CA model. Although the study area is not highly urbanized, the overall accuracy of our model can achieve to 90.79%. More specifically, my approach has a better performance when identifying non-urban areas to urban areas (accuracy of CNN-CA model is 78%, whereas the accuracy of logistic regression-CNN model is 2.4%).

While the approach in this study presents the ability of deep learning techniques and the practicability of spatially explicit hyperparameter optimization, future work can concentrate on the following aspects. First, I will apply the CNN-CA model to other study regions and compare the local-level (county-based) accuracy and global-level accuracy (entire study area). Second, I will adopt agent-based model in order to further investigate the land change process. Third, I will further address the computational challenge for big data-driven CNN model.

References

Bai, J.H., Hua Ouyang, Zhifeng Yang, Baoshan Cui, Lijuan Cui, and Qinggai Wang. 2005. Changes in wetland landscape patterns: A review. *Progress in Geography* 24 (4): 36–45.

Batty, Michael, Helen Couclelis, and Mark Eichen. 1997. *Urban systems as cellular automata.* London, England: SAGE Publications, Sage UK.

Batty, Michael, and Yichun Xie. 1994. From cells to cities. *Environment Planning B: Planning Design* 21 (7): S31–S48.

Bengio, Yoshua. 2012. "Practical recommendations for gradient-based training of deep architectures." In *Neural networks: Tricks of the trade*, 437–478. Springer.

Biau, GÃŠrard. 2012. Analysis of a random forests model. *Journal of Machine Learning Research* 13 (Apr): 1063–1095.

Castle, Christian JE, and Andrew T Crooks. 2006. Principles and concepts of agent-based modelling for developing geospatial simulations.

Choi, Samuel PM., Jiming Liu, and Sheung-Ping. Chan. 2001. A genetic agent-based negotiation system. *Computer Networks* 37 (2): 195–204.

Chollet, François. 2017. Xception: Deep learning with depthwise separable convolutions. In *Proceedings of the IEEE conference on computer vision and pattern recognition.*

Du, Jingcheng, Lu Tang, Yang Xiang, Degui Zhi, Jun Xu, Hsing-Yi Song, and Cui Tao. 2018. Public perception analysis of tweets during the 2015 measles outbreak: comparative study using convolutional neural network models. *Journal of Medical Internet Research* 20 (7): e236.

Fielding, Alan H., and John F. Bell. 1997. A review of methods for the assessment of prediction errors in conservation presence/absence models. *Environmental Conservation* 24 (1): 38–49.

Goodfellow, Ian, Yoshua Bengio, and Aaron Courville. 2016. *Deep learning*: MIT press.

He, Jialv, Xia Li, Yao Yao, Ye Hong, and Zhang Jinbao. 2018. Mining transition rules of cellular automata for simulating urban expansion by using the deep learning techniques. *International Journal of Geographical Information Science* 32 (10): 2076–2097.

Hinton, Geoffrey E, Nitish Srivastava, Alex Krizhevsky, Ilya Sutskever, and Ruslan R Salakhutdinov. 2012. Improving neural networks by preventing co-adaptation of feature detectors. *arXiv preprint* arXiv:1207.0580.

Jean, Neal, Marshall Burke, W. Michael Xie, Matthew Davis, David B. Lobell, and Stefano Ermon. 2016. Combining satellite imagery and machine learning to predict poverty. *Science* 353 (6301): 790–794.

Krizhevsky, Alex, Ilya Sutskever, and Geoffrey E Hinton. 2012. Imagenet classification with deep convolutional neural networks. Advances in neural information processing systems.

Kuhn, Max, and Kjell Johnson. 2013. *Applied predictive modeling*, vol. 26: Springer.

LeCun, Yann, Yoshua Bengio, and Geoffrey Hinton. 2015. Deep learning. *Nature* 521 (7553): 436–444.

Li, H, and JF Reynolds. 1997. Modeling effects of spatial pattern, drought, and grazing on rates of rangeland degradation: A combined Markov and cellular automaton approach. *Scale in Remote Sensing and GIS*: 211–230.

Li, Xia, Jinyao Lin, Yimin Chen, Xiaoping Liu, and Bin Ai. 2013. Calibrating cellular automata based on landscape metrics by using genetic algorithms. *International Journal of Geographical Information Science* 27 (3): 594–613.

Li, Xia, Qingsheng Yang, and Xiaoping Liu. 2008. Discovering and evaluating urban signatures for simulating compact development using cellular automata. *Landscape and Urban Planning* 86 (2): 177–186.

Li, Xia, and Anthony Gar-On Yeh. 2000. Modelling sustainable urban development by the integration of constrained cellular automata and GIS. *International Journal of Geographical Information Science* 14 (2):131–152.

Li, Xia, and Anthony Gar-On Yeh. 2002. Neural-network-based cellular automata for simulating multiple land use changes using GIS. *International Journal of Geographical Information Science* 16 (4): 323–343.

Liang, Xun, Xiaoping Liu, Xia Li, Yimin Chen, He Tian, and Yao Yao. 2018. Delineating multi-scenario urban growth boundaries with a CA-based FLUS model and morphological method. *Landscape and Urban Planning* 177: 47–63.

Lin, Jinyao, and Xia Li. 2015. Simulating urban growth in a metropolitan area based on weighted urban flows by using web search engine. *International Journal of Geographical Information Science* 29 (10): 1721–1736.

Pontius, Robert Gilmore, Wideke Boersma, Jean-Christophe. Castella, Keith Clarke, Ton de Nijs, Charles Dietzel, Zengqiang Duan, Eric Fotsing, Noah Goldstein, and Kasper Kok. 2008. Comparing the input, output, and validation maps for several models of land change. *The Annals of Regional Science* 42 (1): 11–37.

Reed, Russell, and Robert J MarksII. 1999. *Neural smithing: Supervised learning in feedforward artificial neural networks*: Mit Press.

Salakhutdinov, Ruslan, and Andriy Mnih. 2008. Bayesian probabilistic matrix factorization using Markov chain Monte Carlo. In *Proceedings of the 25th international conference on Machine learning*.

Schmidhuber, Jürgen. 2015. Deep learning in neural networks: An overview. *Neural Networks* 61: 85–117.

Srivastava, Nitish, Geoffrey Hinton, Alex Krizhevsky, Ilya Sutskever, and Ruslan Salakhutdinov. 2014. Dropout: A simple way to prevent neural networks from overfitting. *The Journal of Machine Learning Research* 15 (1): 1929–1958.

Tang, Wenwu, and Jianxin Yang. 2020. Agent-based land change modeling of a large watershed: Space-time locations of critical threshold. *Journal of Artificial Societies and Social Simulation* 23 (1): 1–15.

Tobler, Waldo R. 1970. A computer movie simulating urban growth in the Detroit region. *Economic Geography* 46 (sup1): 234–240.

UNDESA. 2018. World urbanization prospects 2018.

Wang, Yeqiao, and Xinsheng Zhang. 2001. A dynamic modeling approach to simulating socioeconomic effects on landscape changes. *Ecological Modelling* 140 (1–2): 141–162.

Wolfram, Stephen. 1993. Statistical mechanics of cellular automata. *Journal Reviews of Modern Physics* 55 (3): 601.

Xiao, Ningchuan, David A. Bennett, and Marc P. Armstrong. 2002. Using evolutionary algorithms to generate alternatives for multiobjective site-search problems. *Environment and Planning A* 34 (4): 639–656.

Xiong, Hui Yuan, Yoseph Barash, and Brendan J. Frey. 2011. Bayesian prediction of tissue-regulated splicing using RNA sequence and cellular context. *Bioinformatics* 27 (18): 2554–2562.

Yang, Qingsheng, Xia Li, and Xun Shi. 2008. Cellular automata for simulating land use changes based on support vector machines. *Computers & Geosciences* 34 (6): 592–602.

Yao, Yao, Jinbao Zhang, Ye Hong, Haolin Liang, and Jialv He. 2018. Mapping fine-scale urban housing prices by fusing remotely sensed imagery and social media data. *Transactions in GIS* 22 (2): 561–581.

Zhai, Yaqian, Yao Yao, Qingfeng Guan, Xun Liang, Xia Li, Yongting Pan, Hanqiu Yue, Zehao Yuan, and Jianfeng Zhou. 2020. Simulating urban land use change by integrating a convolutional neural network with vector-based cellular automata. *International Journal of Geographical Information Science*: 1–25.

Zheng, Minrui, Wenwu Tang, and Xiang Zhao. 2019. Hyperparameter optimization of neural network-driven spatial models accelerated using cyber-enabled high-performance computing. *International Journal of Geographical Information Science*. https://doi.org/10.1080/13658816.2018.1530355

Chapter 7
Conclusion

The primary purpose of this book is to propose a framework that could apply to machine learning algorithm-based spatial models with consideration of spatial features. The integration of machine learning and GISscience (the central topic in this book) provides support for improving current hyperparameter optimization approaches. The spatially explicit hyperparameter optimization approach proposed in here focuses on achieving three research objectives: (1) examining the feasibility and necessity of spatially explicit hyperparameter optimization in spatial models, (2) addressing the computational efficiency issue from model- and computing-level, and (3) investigating the practicability of spatially explicit hyperparameter optimization in different spatial models. To author's knowledge, this study is the first to integrate methods from GIScience with conventional hyperparameter optimization approaches from computer science field.

In Chap. 4, I proposed a spatially explicit hyperparameter optimization approach for neural network-based spatial models. In this approach, I incorporated methods from GIScience (e.g., spatial statistics) to explore the local variation structure of the search space of hyperparameters, and further to adjust the local variation structure based on spatial dependence. Further, methods from GIScience (i.e., spatial sampling and second-phase sampling) addressed the computational-intensity issue from the model level. This conclusion links to objective 1.

Chapter 5 (corresponding to objective 2) discussed the automated framework of spatially explicit hyperparameter optimization. This framework has three components: automatic search of hyperparameters, spatial prediction of hyperparameter space, and acceleration of hyperparameter search. The results demonstrate that spatially explicit hyperparameter optimization can excavate the landscape of hyperparameters space and adjust the local landscape with spatial dependence. Also, spatially explicit hyperparameter optimization significantly accelerates the search process of EA (about 100 generations down to about 40 generations).

In order to examine the practicability of spatially explicit hyperparameter optimization (links to objective 3), I applied this approach in a CNN-CA model. The results from Chap. 6 indicate that spatially explicit hyperparameter optimization

M. Zheng, *Spatially Explicit Hyperparameter Optimization for Neural Networks*,
https://doi.org/10.1007/978-981-16-5399-5_7

approach could find appropriate hyperparameters for CNN model in a short time. The CNN-CA model with appropriate hyperparameters has a better performance than logistic regression-CA model. The overall accuracy can achieve to 90.79%.

The research presented here contributes to the GIScience community in several ways. First, spatially explicit hyperparameter optimization is an important addition to the field of GIScience. The importance of hyperparameters for improving the performance of machine learning algorithms has been discussed in the computer science field for some time. Spatially explicit hyperparameter optimization opens a new world for researchers to use appropriate hyperparameters to improve model performance in GIScience. Second, spatially explicit hyperparameter optimization fills the gap where there is no hyperparameter optimization approach considering spatial dependence in the search space. Also, the approach proposed in this book can explore the landscape of the search space of hyperparameters. Last but not least, spatially explicit hyperparameter optimization further reduces computational burden at both model and computing levels.

Of course, each method holds its own limitations and has potential room for improvement. Based on current work, the future study should explore the following areas: (1) integration of a stand-alone application that covers all components of spatial explicit hyperparameter optimization; (2) comparison of the current framework with the latest population-based hyperparameter optimization, such as PBT (Population Based Training) for neural network; (3) optimization of the current evolutionary algorithm; (4) automate generation of the landscape map of the search space; and (5) addition of user-defined components into current framework, such as accuracy metrics.

Printed in the United States
by Baker & Taylor Publisher Services